D0329172

BELMONT UNIVERSITY LIBRARY
BELMONT UNIVERSITY
1900 BELMONT BLVD.
NASHVILLE, TN 37212

AIR POLLUTION METEOROLOGY

"Talking of education, people have now-a-days" (said he) "got a strange opinion that every thing should be taught by lectures. Now, I cannot see that lectures can do so much good as reading the books from which the lectures are taken. I know nothing that can be best taught by lectures, except where experiments are to be shewn. You may teach chymistry by lectures – you might teach making of shoes by lectures!"

James Boswell: *Life of Samuel Johnson, 1766*

Dedicated to my wife Margaret: *sine qua non*

ABOUT OUR AUTHOR

Richard Scorer gained an open scholarship to Repton School, and another to Cambridge University in Mathematics where he gained the title of Wrangler for Part II Mathematics examination. He obtained his PhD at Cambridge for a doctorial thesis on Atmospheric Waves which answered a problem posed by an earlier air crash in North Wales. He moved to Imperial College, London where he has since been Lecturer in Meteorology, Reader in Applied Mathematics, and Professor of Theoretical Mechanics; and is currently Emeritus Professor.

The following catalogue of academic appointments and public service makes impressive reading:

Founding member of the journal *Atmospheric Environment*; Fellow of the Royal Society of Health; Fellow of the Institute of Mathematics and its Applications; Fellow of the Royal Aeronautical Society; Honorary Member and former Chairman of the National Society for Clean Air and Environmental Protection (NSCA); Honorary Member of the Czechoslovak Meteorological Society; Member of the American Association for the Advancement of Science; Fellow of the Royal Meteorological Society; Former President and Honorary Member of the Royal Meteorological Society; Advisor on environmental research to the Central Electricity Research Laboratory; Member and former Committee Chairman of the Technical Committee of the UK Clean Air Council; Member of the Farm and Food Society; Conservator of Wimbledon Common.

Since retiring, Professor Scorer has spent more time on the new technology of satellite imagery of clouds. He has written many books on meteorology and aerodynamics which are all leaders in the field, especially for the understanding of cloud mechanisms in the atmosphere. He has travelled widely all over the world, lately in the Gallapagos Islands; Spitsbergen, Iceland, Greenland; South Georgia and the Arctic Peninsular, as well as examining the far North during his satellite studies, and has always been very aware of the ecological and environmental forces at work on all life forms on Earth. These travels have acutely impressed upon the author the magnitude of the human influence on the welfare of other species and flora of the world, and their influence on us.

AIR POLLUTION METEOROLOGY

Richard S. Scorer
Emeritus Professor and Senior Research Fellow
Imperial College, London

Horwood Publishing
Chichester

BELMONT UNIVERSITY LIBRARY

First published in 2002 by
HORWOOD PUBLISHING LIMITED
International Publishers in Science and Technology
Coll House, Westergate, Chichester, West Sussex PO20 6QL
England

COPYRIGHT NOTICE
All Rights Reserved. No part of this publication may be reproduced, stored in a retrieval system, or transmitted, in any form or by any means, electronic, mechanical, photocopying, recording, or otherwise, without the permission of Horwood Publishing, Coll House, Westergate, Chichester, West Sussex, England.

© R.S. Scorer, 2002

British Library Cataloguing in Publication Data
A catalogue record of this book is available from the British Library

ISBN: 1-898563-93-4

Printed & bound by Antony Rowe Ltd, Eastbourne

BELMONT UNIVERSITY LIBRARY

QC
882
.B36
2002

Author's Preface

Meteorology is a study of how Nature produces such a glorious variety of weather and its advance gradually makes forecasting possible. We have come to regard the atmosphere as a fluid and understanding it requires making mathematical models which must obey the laws of fluid mechanics and physics. Professor David Brunt used to say that the air is basically water vapour diluted with large amounts of oxygen and nitrogen which changed very little while the water vapour condensed into clouds which collected most of the pollution and precipitated it in rain, snow, hail, and thereby produced the continual display of beautiful, severe and very varied weather.

Today the various gases which complete a roll-call of vast chemical variety, while all the different parts of the atmosphere transform them from primary to secondary pollution which then plays tricks with sunshine and radiations far away in the ultra-violet and infra red. So we are interested already in the consequences that will follow the release of pollution into the weather systems which will come tomorrow and beyond.

We start by describing what this pollution is, for it is mostly the result of burning fossil fuel and great industries manufacture all sorts of chemicals which react with sunshine over a very wide range of temperatures. We already have many books describing the clouds and storms in the process of routine weather forecasting using a wide range of mathematical models made possible by modern computers which can carry and process the millions of observations which we feed into them.

As we do this the scientific problems of physics and chemistry are multiplied. They are thrown around by the mechanics of the fluid atmosphere and the radiation from the sun. We may hope, or even assume, that it is merely a question of processing the observations through suitable computers to obtain a prediction of tomorrow's pollution display. But we have to face that any such forecast must follow, not precede, the forecast of the weather. And we know that the weather can drift apart from the forecast often by tomorrow. So do we need to take the laws of physics literally and in details?

To explore these possibilities we study the kind of meteorology which air pollution tells us about, and how it may be transformed.

This book records how the subject has changed and is changing, in the hope that the mistakes of the (recent) past will not be followed again.

The terminology of meteorology becomes so familiar over the years that it seems simple to an old hand. Therefore this time I have included a technical dictionary, which is more than an enlarged glossary; it supplements many aspects of the earlier chapters, and it should be helpful to budding meteorologists, and I hope it helps them to get more from what they can see with their otherwise unaided eyes.

Everyone concerned with air pollution thinks, or even knows, that it is dispersed by turbulence; but what is turbulence? Read on

Acknowledgement

I am deeply indebted to my daughter Beatrice who salvaged the text from my old computer when it gave up, and compiled the camera-ready copy for this book from heaps of text, slides and photographs. I am very grateful to her for taking on this task while also preparing to move to work in Uganda, and for completing it before leaving the country!

Any remaining errors are my responsibility.

R.S.Scorer
Raynes Park
2002

Table of Contents

1. The meteorological scene 1
    1.1   Gases and clouds 1
    1.2   Grit and dust 1
    1.3   The Planck curve 3
    1.4   The air temperature profile 5
    1.5   The ozone layer 7
    1.6   Stratospheric chemistry 8
    1.7   Cloud trails 10
    1.8   Stratospheric wave clouds 12
    1.9   Global warming by carbon dioxide 13
    1.10 Greenhouse gases 14
    1.11 Climate change 16
    1.12 Visual Observation 16
    1.13 Trade wind convection 17
    1.14 Stability, convection, lapse-rate 18

2. Chimneys: purpose and design 21
    2.1   The concept of a plume 21
    2.2   The two and a half times rule 23
    2.3   Flagging 27
    2.4   Strong winds 28
    2.5   Stack oscillation in the wind 30
    2.6   Thermal rise and effective stack height 31
    2.7   The problem of several flues close together 34
    2.8   The advantage of a multi-flue stack 36
    2.9   Bifurcation of a buoyant plume 38

3. Dispersal in theory and practice 43
    3.1   Inevitable inaccuracies 43
    3.2   Models and legislation 44
    3.3   Dispersal of a solitary plume: theory 47
    3.4   Maximum ground level concentration (max GLC) ... 51
    3.5   The mixing layer 54
    3.6   The sub-cloud inversion 56
    3.7   Non-dispersing plumes aloft 59
    3.8   Non-dispersing plumes at the surface 61
    3.9   Theoretical second maximum at the ground 62

3.10   Comment: the theoretical bent-over plume 63
3.11   The nature of actual turbulence 66

4. Acid rain and development of gas washing 69
    4.1   Acid rain, and the health of fresh water fish 69
    4.2   The Loch Fleet study 72
    4.3   The concentration of acidity 72
    4.4   The Battersea saga 73
    4.5   Different features of the Bankside plume 78
    4.6   Avoidance of cold effluent: flue gas desulphurisation (FGD) 81

5. Environmental costs and benefits: a matrix for comparison of sources 83
    5.1   The method 83
    5.2   Matrix details 85
    5.3   Numerical answers 88
    5.4   Acid gases: acid rain 91
    5.5   New problems: straw burning, background pollution 94
    5.6   Blackening of surfaces: traffic exhaust 97
    5.7   The PM_{10} problem 101
    5.8   The PM_{10} problem turns medical, but what is its priority? 103
    5.9   The Utah valley steel mill 104

6. The Technical Dictionary 107
    Index of the Technical Dictionary 145

References 147

Index 149

1

The meteorological scene

1.1 GASES AND CLOUDS

Air pollution is composed of gases or clouds of particles which cause us a nuisance of some sort. There are certain natural phenomena which are often in the same category as pollution which may be a useful guide to our thinking about pollution because they have to follow the same laws of mechanics and physics as the pollution.

Condensation of water vapour into fog and clouds produces condensation droplets, or particles, which are so small that they are carried by the motion of the air and may be described as 'floating' in the air. But it is not buoyancy that prevents them from falling towards the ground because they are mostly heavier than the air they displace and it is the viscosity of the air that reduces their fall speed so that it is small compared with the motion of the air which may be very turbulent at times.

Smoke is also a condensation of vapour (or vapours) which are boiled off the fuel. That is why combustion experts always describe smoke as wasted fuel which ought to be burnt in the fire. The stove, or cylinder, ought to be designed to cause this to happen.

Ash is the other main product of burning fuel or of incinerating rubbish of various kinds. Bonfires make a great distinction between smoke, which drifts with the wind, and ash, which mostly remains in the fireplace. The designers of most industrial furnaces and heating systems have tried to produce a controllable draught of air which would facilitate the burning of all that was combustible. Some of the ash may unavoidably be carried out of the chimney top, especially where a strong draught is used with pulverised solid fuel. This ash is small-sized grit and is deposited slowly over a fairly large area of countryside. It may consist of small hollow spheres from power stations or cement works when pulverised coal is used as the fuel.

1.2 GRIT AND DUST

These are very ordinary, for they are often raised up from the ground by the wind. *Grit* is a collection of the larger particles in this category and therefore falls to the surface close to its place of origin. *Dust* may be carried much further; perhaps even as far as the icy wastes around the North or South pole.

Dust is easily seen because it is usually present as clouds in which the particles descend towards the surface only slowly. *Grit*, by contrast, falls only too rapidly and the particles are separated so far apart that, being too small to see individually, they do not obscure the vision like a cloud but are, nevertheless much more likely to hit you in the eye.

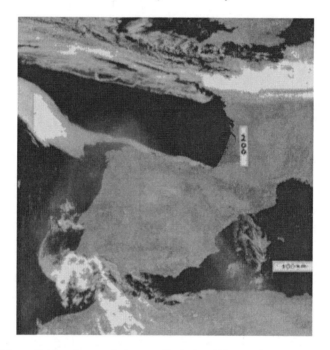

1.2.1   Sahara dust carried across Spain into Biscay where we meet air from the far North detected by the ship trails which show that it has been starved of CCN by resting over ice, and then sprinkled with ship exhaust. *Picture time 11.40 UT, date 21.08.80 Univ. of Dundee (satellite CZCS)*

The ordinary picture of the Sahara, the Gobi, or any other big desert is of sand dunes and dust storms. On Mars the surface is so dry, and has been for millions of years, that wind storms tend to fill the air with dust so that the surface is then not visible from outside. The same is true of the big deserts on Earth and the wind which blows up dust clouds is often started by the downdraught of a thunderstorm. The squall which blows up a 'haboob' is caused by the arrival at the ground of the cold air brought down by evaporating rain. After the squall has passed by, the cold air moves onto a hot, sandy surface which starts strong buoyant convection in the cooler air.

1.2.2   Gobi dust which has crossed North China and is carried out to sea and across the South of Japan. *05.33, 15.4.79 vis GMSI Japan Meteorological agency*

1.2.3   A haboob in the Sudan. *Crown copyright*

1.3   THE PLANCK CURVE

Over the ocean a very hazy air mass is sometimes formed. This may be due to a strong wind blowing up sea spray. If the air is rather dry the bubbles of spray become shattered into a very large number of tiny sea salt particles. If the air then moves to a cooler region the relative humidity is increased as the air is cooled to be nearer its dew point, and the particles become larger because the sea salt is hygroscopic (and combines with water vapour). The result is that the visibility is reduced, and people may be prompted to ask where the 'pollution' has come from. Genuine sea fog may arise, typically around the coasts of north-west Europe when radiation fog drifts off the coast and is then drawn back towards the land by the onset of sea breezes.

We know instinctively that pollution will be a worse nuisance when there is little or no wind to 'carry it away'. Today it is common knowledge that some pollutants which do not occur naturally are beginning to show a gradual increase everywhere, which indicates that there is no sufficiently well known natural mechanism to remove them from the atmosphere. The most notorious are the ChloroFluoroCarbons (CFCs), which were becoming useful and valuable because of their chemical properties. The only way they could be destroyed or otherwise removed from the troposphere would be by being carried up into the stratosphere and decomposed by the ultra-violet (UV) component of sunshine, which is too feeble to do that at the surface.

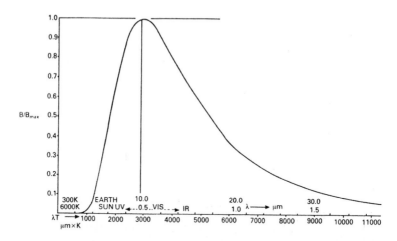

1.3.1   The Planck curve showing how the radiation emitted depends on the product of the temperature and the wavelength shown on the scale at intervals of 1000 units for the case of the Sun at 6000 K. On the line above the scale for the Earth at 300 K is shown. On the scale for the sun the positions of the UV, the visible, and the IR are shown on a wavelength scale.

The range of the **wavelengths** (in microns) of the **ultra-violet (UV)** is less than 0.4 μm; of the **visible (vis)** is 0.4 to 0.7 μm; and of the **infra-red, (IR)** is greater than 0.7 μm as indicated in figure 1.3.1. The maximum intensity of sunshine is close to 0.5 μm in the blue-green region. The sunshine at wavelength less than 0.3 μm is call UVB

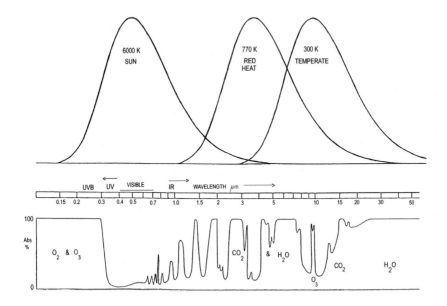

1.3.2   This diagram shows the energy spectrum of radiation emitted by bodies at three different temperatures with amplitude unity at each maximum. The middle scale shows the wavelength and is logarithmic which is why the emission curves are more symmetrical than in figure 1.3.1. The bottom curve shows the absorption by the troposphere. It is readily seen that the visible range of wavelengths is not emitted at all by bodies cooler than about 700 K. The radiation emitted by the Earth is largely absorbed in the atmosphere, mainly by water vapour and carbon dioxide. The incoming sunshine, which is largely in the visible range of wavelengths, is much less absorbed by the atmosphere and so most of its energy reaches to clouds or the ground where it is either absorbed or reflected according to the colour of the surface.

1.4   THE AIR TEMPERATURE PROFILE

In the layers where sunshine or some other heat source provides warmth at the bottom of the atmosphere the buoyancy makes it unstable where the air is cooler aloft and convection takes place. Thus is formed the **mixing layer** which contains the point of entry of much pollution.

As it rises upwards to lower pressure it becomes cooler and at the **condensation level** cloud composed of water droplets is formed. This releases latent heat and this reduces the rate of cooling in rising air from the 'dry'- to the 'saturated'-**adiabatic lapse rate** and also produces the **sub-cloud inversion** close to that transition.

The layer in which clouds and storms produce the weather is the **troposphere** which extends up to around 18 km or so in the tropics but only up to 9 to 12 km in temperate latitudes and is limited to only 3 to 6 km in polar regions. The top of that layer is called the **tropopause**, and over the great ice plateaux of Greenland and Antarctica it may be as low as the snow surface. Above the tropopause the air is stably stratified for about the next 20 km or so with a fairly constant temperature. In the troposphere the temperature is mainly changed by the vertical motion; but from the tropopause level upwards radiative exchanges

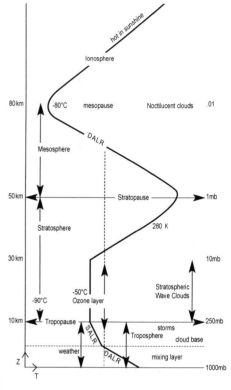

with the total environment from the land and sea to outer space begin to be dominant, and it becomes warmer reaching a near-surface temperature at the **Stratopause** at around 290 K at a height of about 50 km. But then outer space begins to take over in the **mesosphere** and by a height of about 80 km at the **mesopause** it has become as cold as −80°C and clouds of what are thought to be ice or water particles, called noctilucent clouds are sometimes seen. They cannot be seen by day because of the glow of the troposphere although there is a brief period after sunset or before sunrise when they could be visible if present.

From the mesospause upwards the air is so rarefied that the mean-free-path of the molecules soon approaches 1 km and the air becomes like a very viscous fluid with a large conductivity for heat. It is appropriately called the **ionosphere** and in so far as the temperature has any precise meaning it steadily increases up to several hundred degrees at greater altitudes due to direct sunshine.

1.5 THE OZONE LAYER

The warmer part of the stratosphere and the ionosphere were discovered during the early part of the 20th century by the fact that sound and radio waves were reflected back down to the ground.

It had already been suggested that the oxides of nitrogen (collectively denoted by NO_x) are produced to some extent in all combustion processes using air to provide the oxygen required. Aero-engines, operating in the stratosphere, might become important factors in the photochemical reactions in the stratosphere. New reactions due to the ultra-violet (UV) component present might upset the existing reactions and the equilibrium of the ozone, which is both created and destroyed by photochemical reactions in the stratosphere by short wave sunshine.

The **ozone layer** is the name given to the levels between about 17 and 24 km. This is the region where the ozone is densest and the UVB is found to be mostly absorbed from the sunshine so that life forms of many different kinds living in the troposphere have been able to evolve without suffering serious sun-burn. Our pale human skins become coloured brown by sunshine but that is due the UVA which is the ultra-violet component of sunshine of longer wavelength than about 3 microns, but the component between 3 and 4 µm is still invisible but does not cause as serious cancers as the UVB. The ozone layer absorbs most of the UVB and it was feared (as a result of taking primitive theories too literally) that the NO_x might cause reactions with UVB which might reduce the ozone (O_3) to O_2 which does not absorb UVB as readily. The scare lasted until it was appreciated that NO_x is already present as one of the gaseous components so that adding a bit more by flying aircraft in the lower stratosphere would probably not much upset things. Commercial air liners already fly above the tropopause because it provides less bumpy flying. Nevertheless aircraft exhaust transports a vast amount of water vapour to the stratosphere.

This is a case where the **precautionary principle** (see chapter 6) began to be invoked because of the uncertainty of the rates at which the various photochemical reactions would take place in the conditions of the stratosphere with very low temperature and low pressure. These conditions could be reproduced in a laboratory only in a refrigerator at reduced pressure, and there was a very serious objection to experiments under such conditions because the reactions to be measured would be affected by the presence of the walls of the refrigerated chamber. At that time (mostly before 1970) frank guesses were made of the reaction rates of the photochemical reactions in the stratosphere, and because of the newness of the subject the guesses, which were mostly extrapolations from existing data, made by the very few experts, could not be subjected to informed criticism as they lacked precision.

1.6   STRATOSPHERIC CHEMISTRY

1.6.1   The tropopause can very often be seen clearly as a sharp inversion at a haze or cloud top. When the cloud is thin the Sun's rays pass through with only one encounter with an ice crystal which therefore causes mainly forward scatter. The cloud is therefore seen from above mainly by the light scattered up by the Earth's surface and lower clouds which is a purplish brown mixture. The low cloud in this view; having very numerous very small water particles appears clean white. The special interest in this case is that the troposphere absorbs many wave bands as shown in 1.3.2 and anything in the stratosphere is subject to intense ultraviolet, UV, radiation from the sun.

It was necessary to make many other guesses because the world-wide distribution of NO_x in the stratosphere was scarcely known and the design of combustion chambers had not yet been studied by the fuel-burning industries to determine what control of NO_x in aero-engines and power stations could be achieved. Additional guesses had to be made concerning the probable, and the possible, extent of aircraft operations in the stratosphere. While that part of the science was being argued about it would also be required to know how rapidly the fluorocarbons and oxides of nitrogen would be carried up naturally into the stratosphere. Indeed it was argued that the transport of NO_x upward across the tropopause naturally was minimal because they were normally very reactive chemically and soluble in the water of cloud droplets so that they would be mostly precipitated out of the troposphere by rain. There were temporary excitements concerning the contribution of nitrous oxide (N_2O) emitted from farmland which was increased by the use of nitrogenous fertilizers. But the panic, caused by the suggestion that the chemistry of ozone in the stratosphere would be completely upset, led to an agreement among governments that the manufacture of chlorofluorocarbons (CFCs) should be ended as soon as possible. Many had regarded it as important to apply what later became known as the *precautionary principle* as rigidly as possible. It was even thought that the importance of this agreement was that the governments had reached agreement at all, and it was taken as a good sign for the inevitable attempt to get agreement to reduce the output of CO_2.

The outcome has been that scientists have been induced to predict calamities and so-called science correspondents of the press have, with pretended authority sometimes, issued absurd versions of these predictions. The public has oscillated between scepticism that doom was being ignorantly forecast and the other extreme that the 'truth' could be 'discovered' by courts of law, parliamentary discussion and vote, or public opinion surveys.

The result has been a growing distrust of science. Yet the outcome has been that observational systems have been greatly improved. For some time no reductions in the amount of ozone in the stratosphere were found and more recently the number of supersonic aircraft flying extensively in the stratosphere had become definitely quite trivial.

The scene was transformed by the discovery of the so-called **Ozone Hole** (see Chapter 4), which subsequent mapping showed to exist more or less over the continental bounds of Antarctica. It has been explained as being due to the existence of wave clouds in the stratosphere caused by the airflow in the south polar vortex flowing over the mountains of Antarctica as it develops each winter. The particles in these clouds are the site on which the photochemical reactions take place at a lower temperature than on any other clouds on Earth. There Chlorine plays a role rather similar to that of NO in taking on another atom of oxygen to become NO_2: atomic chlorine becomes ClO. Since UV energy is required to take the molecules through the sequence by which ozone (O_3) becomes oxygen (O_2), the Antarctic store of ozone, which had been known normally to last through the southern winter, begins to be destroyed in October with the onset of the Southern spring; and this produces an annual global deficit of ozone for the year.

1.7   CLOUD TRAILS

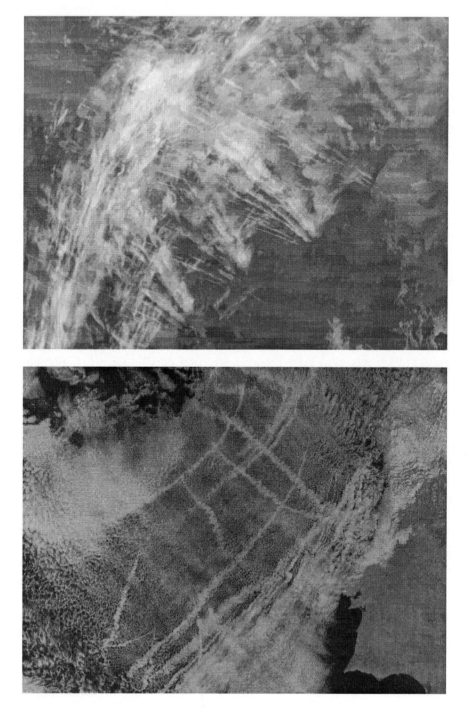

Aircraft Condensation trails may persist for several hours provided that they are frozen immediately after formation. The most persistent are close to the tropopause. The trails are composed of water vapour together with condensation nuclei which are produced in very large numbers in almost any combustion process. To remain visible the droplets must be frozen before they evaporate, which means that they must be in air with a temperature very close to, or more probably below −40°C. Brightly coloured mock suns are commonly seen at 22° from the sun, and at the sun's altitude, in contrails, which indicates that ice crystals are hexagonal prisms with vertical axes. The amount of cloud created in this way has very little effect on the radiation balance by day, but may have a small warming effect if they have been spread by wind shear, at night. Trails provide some insight into the dispersion mechanisms at high altitudes.

1.7.1   (opposite) As a general rule, **contrails** persist for several hours provided they are frozen immediately after formation. They are formed by condensation of water vapour on the very numerous condensation nuclei (CCN) both of which are supplied by the aircraft exhaust. Successive aircraft use the same navigation point. *NOAA 10, 19.01; 28.08.87, chan4 57N 10W Univ. Dundee*

1.7.2   (opposite) By contrast there is no deficiency of water vapour close to the sea surface, but if the air has been in a region where the Cloud condensation nuclei (CCN) fall out over time (e.g. the polar North), **shiptrails**, which are much whiter than the surrounding cloud, are formed on the cloud condensation nuclei (CCN) in ship's exhaust. *NOAA 7, 16.24; 17.08.83 chan2 60N 32W Univ. Dundee*

The frequent occurrence of large areas of cloud produced artificially has given rise to the suggestion that their creation might be used to produce a bit of 'global cooling'! This could then be an antidote to the global warming which seems regrettably to be taking place. And it raises the question of the cost of mounting the necessary operations to produce an appreciable effect.

Although contrails are formed on almost any day, the same is not true of ship trails which are limited to only one or two days in three weeks, and even then take a long time to produce and are rather restricted to parts of the Atlantic and Pacific oceans.

There has been no useful estimate of the average duration of either type of trail. This could be very important because the contribution to warming or cooling is different according to whether it is day or night; and either kind of trail may last more than 12, or even 24 hours.

Nor would we have any method of cancelling out either kind to prevent them lasting beyond sunrise or sunset when they look likely to cancel out any useful work already done.

It may be remarked that the introduction of a form of pollution to cancel out some that is already producing an unwanted effect is unlikely to be helpful if only because they are all being diluted all the time. The only remedy for any unwanted pollution is to avoid its emission in the first place.

One of the big problems for chemists is what happens in the stratosphere in UV. Basic to any modelling of these processes are.

(1)   the chemistry at 100 mb and a temperature of $-90°C$, both difficult to model in the laboratory; and

(2)   to estimate how well the aircraft exhaust is mixed with the surroundings. The indications are, from the persistence of contrails, that the mixing is very slow.

It is also noteworthy in this connection that where detailed observations of the atmosphere's structure are made it is found that there are many layers which are not mixed, and have thicknesses which are as small as can be measured. What is needed, for producing the photochemical reactions is some sort of mixing and as would be found in the troposphere in the form of cloud crystals or drops which have a not-negligible fall speed at those levels and have a mixed wake. It has indeed been suggested that the particles in stratospheric wave clouds do indeed fulfil this role. A few observations of the composition of these wave cloud particles must be obtained before numbers are put to any predictions about photochemical reactions.

1.8   STRATOSPHERIC WAVE CLOUDS

A mechanism, which operates in stratospheric wave clouds (see Chapter 4) and produces the *ozone hole* over the high mountains of Antarctica in the south, also operates in stratospheric wave clouds in the far North but to a much less extent. There are clear causes for this difference: the North pole is surrounded by ocean whereas the South Pole is surrounded by a continent a large proportion of which is at a height of 2 km or more above sea level, and so the surface beneath the atmosphere is mostly 20°C colder in the South, and there there is less low cloud so that the surface loses more heat by radiation to the sky and such clouds as form in the stratosphere often have a lower equilibrium temperature (below $-90°C$) in the South. The principal wave clouds over mountains are formed much further away from the pole in the Arctic North.

Wave clouds do form in the North (around -80 to $-85°C$) over Alaska, Baffin Island, Ellesmere Island, Greenland, Iceland, Northern Scandinavia, and parts of Northern Siberia. But the winter circumpolar vortex is believed to be much less constant in position and duration than over Antarctica.

Consequently the destruction of the CFCs and other chlorine-based molecules by UV sunshine in the stratosphere and their absorption in the acid droplets of the wave clouds and contrails at temperatures colder than $-90°C$ is far less extensive in the North. The destruction of ozone in UV sunshine in the presence of supercooled acid droplets is relatively rare in the North. To complete the required picture we note that the ozone and the cloud particles move along with the wind; with the droplets forming at the upwind edge of the cloud and evaporating at the downwind side of the cloud; the cloud itself being stationary in a fixed position relative to the mountain which caused it. As the wind system changes the clouds shift around and probably remain in existence longer in the South. We believe that the clouds are often composed of supercooled acid droplets, because of the iridescent colours which can only be produced by spherical particles of fairly uniform size in any region of the cloud. Thus iridescence is common in wave clouds particularly, and to produce these at such cold temperatures without the droplets becoming ice requires them to be acid and remaining supercooled below normal freezing temperature for a few, or more, hours.

1.9   GLOBAL WARMING BY CARBON DIOXIDE

The hard lesson to be learned from the new knowledge about ozone destruction is that the stratosphere was the laboratory in which the mechanisms came to be revealed, not an artificial laboratory cloud chamber refrigerated at very low pressure on the ground.

For any model or prediction of the transport of pollution away from its source a weather forecast is required because the changing situation is always too complicated to include it in a mathematical formula representing the several dispersion mechanisms. Weather forecasts become quite unreliable from this point of view after as little as one day (but rather longer sometimes).

But we do understand how the weather is transformed from summer to winter and back every year. This means that we can predict the trends which vary with the case.

The mechanism of particular interest is the arrival of heat by visible radiation (around 0.5 microns wavelength) from the sun (which is at around 6000 K). At the same time the loss by the Infra-Red radiation of relatively long wavelengths around 10 μm, from the Earth (which is at around 300 K) in all directions to space. These two fluxes must convey about the same amounts of energy in an average year or the Earth would get colder or hotter according to which flux became greater.

This issue was discussed by meteorologists from early in the 19th century when the new industries were pouring great amounts of smoke into the air and obscuring the sunshine and reflecting it back into space. They understood that this would cool the atmosphere. They also realised that perhaps the increase of carbon dioxide caused by the burning of carbon fuels would absorb an equal amount of the outgoing IR from the earth to counterbalance the loss of sunshine. It was thought that if too much was emitted by industry so that the concentration in the atmosphere increased, absorption of CO_2 by the ocean would automatically keep the amount in the atmosphere constant. After all the amount already in the ocean was about 80 times the amount in the atmosphere.

That argument implied that the ocean somehow determined the amount of CO_2 in the air. It became known that this was an error when the figures obtained in Hawaii, isolated in the central Pacific, of the amount of CO_2 in the air revealed that it was altered between summer and winter because there was more vegetation in the northern hemisphere. Nevertheless there was a gradual increase going on at the same time and it has recently reached 400 ppm. It appeared that there was no tendency for the amount in the oceans to have an important role in determining the amount in the air.

1.10   GREENHOUSE GASES

Although carbon dioxide is an important factor in determining the radiative heat loss by the Earth's surface to outer space the water vapour in the air has a similar but actually larger effect. Furthermore there are other gases having molecules composed of 3 or more atoms which are also effective but less so because they are present in smaller amounts.

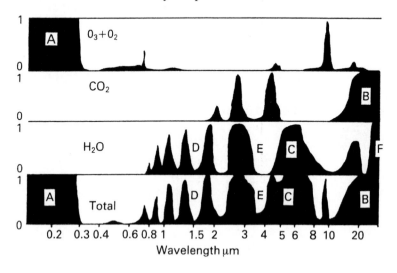

1 10.1   The most important absorbing gases are effective at A, B and C with important windows at D and E

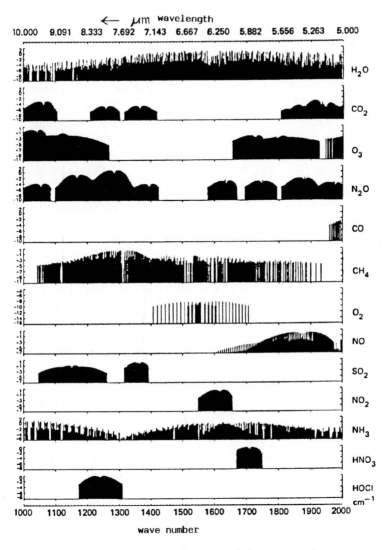

1.10.2   The Greenhouse Gases have very different absorption spectra, as seen here, according to their molecular structure. The important ones have long lifetimes in the atmosphere – water vapour, carbon dioxide, but not so plentiful – methane (CH_4) and its compounds: while others are quickly removed by chemical reaction, e.g. ammonia (NH_3) and sulphur dioxide (SO_2) and therefore do not accumulate in the atmosphere as a whole. The oxides of nitrogen are responsible for much of the acidity in rain together with the oxides of sulphur.

Here are shown more detailed absorption/emission spectra of various major and minor atmospheric gases in the range (IR only) 5–10 µm.

1.11 CLIMATE CHANGE
The possible, and probable, causes of climatic change have characterised the discussions in the 20th century, and because we are observing what seems to be a gradual warming of the Earth's climate the debate has changed to whether the emissions due to industrialisation of mankind are the cause of the change. It is reasonably thought that the little ice age, which began in the 14th century and welcomed in the Black Death, was gradually coming to an end with the passage of the 19th century.

During the 20th century the focus of attention on air pollution has changed from the horrors of pollution in and around our industrial cities to the loading of our whole atmosphere with smokes and gases which may cause climate change, and a very considerable amount of resultant ill health.

The *precautionary principle* appears likely to play a more forceful role than the mechanics of cleaning the air which has taken most of the attention since the end of the second world war.

From now onwards we have to think of planning several decades ahead because climate change is a relatively slow process, and whatever action we take as a result of our thinking about the climate will only take full effect after a few decades, or perhaps even centuries.

Now that we have a better and more extensive array of observing points in operation than in any previous decade we can hope to keep a better eye on the climate, with some hope of anticipating the most dangerous changes.

1.12 VISUAL OBSERVATION
The fluid motion of the atmosphere is mostly invisible, and this has made fluid motion in the laboratory a very attractive idea. Such experiments as have been made have, on the whole, been very helpful in understanding the effects larger in the atmosphere than in the wind tunnel.

Mostly laboratory model experiments illustrate only simple cases and the atmospheric reality is almost always much more complicated. The difficulties mostly originate in the fact that the air in which the weather occurs is not constrained by vertical boundaries and is often stratified in a rather complicated way so that the wind and temperature profiles require more than two parameters to define them and an expensive apparatus to observe and reproduce them in a model. Some people are in excellent positions to observe personally, and this is particularly true of glider pilots. Even their reports of what is going on around them are very much influenced by the theoretical model they have in mind as they make the observations.

It might be supposed that meteorologists would have been able to predict how glider pilots would make the best use of thermals, and this has become normal at gliding competitions. But the early advice was that the cumulus tops had been observed by meteorologists over a few hours in the morning and they had reported that soaring in thermal upcurrents would not be possible: the cloud tops had been observed over hourly intervals and the fastest rising tops were reported to have risen at only around half the sinking speed of the best gliders.

Actually it was only the tops of what the observer had thought were the most rapidly rising ones, and even those had not been measured at intervals of a few seconds at most, but at more like hourly intervals. The great error was that they had not noticed that most individual cumulus only lasted for 15 or 20 minutes or even less. The error was soon cleared up when the aviators discovered real soaring, and the meteorologists had to have a more careful look at the real time scales involved.

But the invisible thermals below cloud were sought out on the assumption that they were warmer in the middle, and pilots went searching for them with a thermometer on each wing tip in the hope that they would then steer towards the warmer air to get the best lift. This idea contained the error that buoyancy would produce a roughly proportional upward velocity, whereas in reality there is no such simple relationship. The upward *acceleration* relative to the immediate surroundings is what the buoyancy produces.

Actuality reveals that a mass of buoyant air by itself forms a ring which, as it rises, causes the air in the middle to rise at twice the speed of the buoyant ring itself (Woodward 1959).

The man who claimed to be able to dissolve clouds by staring very intensely at them had discovered his power because of the intensity of his staring. All he had really discovered was that most cumulus clouds last only a few minutes, and that he had supposed that all the others he wasn't staring at had remained unchanged in the sky (because) he wasn't looking at them; and that was why his staring had to be very intense – to stop him noticing that some of the other clouds were growing larger and others were collapsing and evaporating dissolving and disappearing. Most ordinary people do not have his conceit.

1.13 TRADE WIND CONVECTION

A study of the trade winds can tell us a lot about how the cumulus develops as the ground warms in the morning sunshine because along the direction of the trade wind the sea is becoming progressively warmer, and a model experiment is being performed before our eyes.

1.13.1   Trade wind cumulus are very important clouds in the mechanics of the atmosphere because they take air from middle latitude oceans and convey water vapour upwards from the warm sea as they approach the equator. Thus the total depth of moist air of the troposphere increases and becomes involved in the convection of tropical typhoons and hurricanes and later in frontal cyclones when most of the air returns towards higher latitudes. (*Joanne Simpson*)

1.14   STABILITY, CONVECTION, LAPSE-RATE

In this context convection means **buoyant** (usually **thermal**) convection. It is intuitive that if a fluid (in this case the air) is heated from below it will be set in upward motion. Air rises when the buoyancy force described by Archimedes is applied to it. We sense this happening when cumulus clouds are formed we could feel the instability in the form of bumpiness in passenger aircraft which are usually flown above, or if that is not suitable for the occasion, in between the cumulus clouds but above the level of their base.

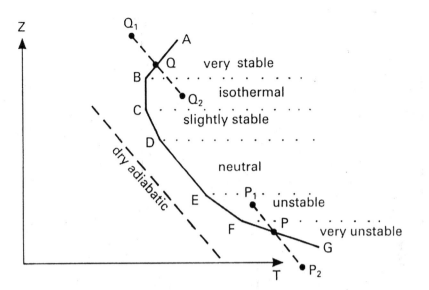

1.14.1   The full curve represents a sounding, i.e. the temperature of the
air as measured. The layer AB represents an inversion layer, and BC an
isothermal layer; both are very stable because a parcel of air would have
to be very hot to rise through them.
CD is a slightly stable layer. In the region AD a displaced parcel would
rise or sink along a line parallel to the dry adiabatic slope, e.g. Q_1Q_2, or
P_1P_2.
DE represents a neutrally stable layer and EF and FG are progressively
more unstable

We also observe that buoyant convection carries air from at or close
to the ground up to the *condensation level* where cloud begins to form. If the
same air rises further more water droplets are formed and then the latent heat
of condensation provides further heat. We can also see that if the cloud
evaporates the effect is to take back the latent heat and cool the air and cause
it to sink. Thus the air between the clouds is not in a state of convection and
is not bumpy to fly through; and so we describe it as being stable - or more
correctly as being **stably stratified**. Such air may also be described as
statically stable, and we call it statically unstable if convection is going on.
In the cumulus situation (of picture 1.13.1) cloudy air is unstable while the
cloud-free air in between is stable, and smooth to fly in.

The movement horizontally in that picture is very important because
it is taking the air westwards and towards the Equator at the same time. This
is one of the Earth's most constant wind systems – the Trade Winds. This is
mentioned because it is also a motion system caused in the first place by the
heating due to sunshine.

The trade winds are composed of air further away from the equator being colder than the air coming from the equator and this trade wind air is being continuously warmed by moving steadily to where the ocean is warmer close to the equator.

Thermal convection consists of the smaller motion systems generated by density gradients due to different heating rates in different places (i.e. at the bottom rather than at the top of an air mass). The larger scale temperature differences between the equator and the poles cause larger scale air motion, which in the context of air pollution we call the wind. We think of it being a constant feature of the situation we may be studying. In that situation convection is an important factor in dispersing pollution into a greater volume of air. 'The wind' carries it 'away'.

That used to be the end of the matter: but no longer, as we shall see.

The vertical temperature gradient determines whether the air is statically stable or unstable. The gradient which divides these two situations is called the *adiabatic lapse rate* which turns out to be a decrease of temperature of very approximately 10°C for an upward displacement of 1 km. This is how much a parcel of air will cool if it is carried up by that distance without any heat coming into or going out of it across its bounding surface. Hence the word adiabatic. Strictly it is required that the parcel does not lose or gain heat by the evaporation of water droplets, and therefore the correct term is the *Dry Adiabatic Lapse Rate* (**DALR**)

If the parcel of air has cloud in it, as it rises it will move to a lower pressure and will cool and condense more cloud and gain some heat thereby and will cool at a reduced rate called the *saturated (or wet for short) adiabatic lapse rate* (**SALR**). In the trade wind cumulus picture (1.13.1) the air is approaching warmer sea nearer the Caribbean. The increasing temperature at the sea surface keeps producing new clouds they rise into drier air, they evaporate a little and this cools the air into which they rise and the lapse rate becomes very slightly steeper than the SALR. In chapter 6 is found the Aerological Diagram in which the lapse rate is defined by the slope of the curve of a sounding, and also a mathematical derivation of the DALR. See also **potential temperature**.

2

Chimneys: purpose and design

2.1 THE CONCEPT OF A PLUME

In this chapter we are concerned with artificial air pollution and the need to render it harmless. Thus we are mainly concerned with the products of combustion, and how to get them taken away.

2.1.1   Smoke from primitive sources (E Africa) *(H. Marchant)*

2.1.2   Pollution accumulated over Lower Hutt, modern suburbs on
Wellington Bay, N.I., New Zealand. (1965)

As it is usually thought that the wind at a high level is stronger than
lower down, an early idea on seeing scenes such as these must be to build
the chimneys as high as possible so as not to allow the accumulation of
pollution in situ at a low level. But there is also a common instinct among
people to spend as little (effort or cash) as possible in getting rid of what is
not wanted, especially if Nature seems to do the job well on most days.

The primitive reaction to smoke produced by the need to cook food
and to keep people warm was to let it out through a hole in the roof of the
living space. The invention of chimneys was to get it away from the
immediate neighbourhood of the hut or tent. Furthermore, the warm air in a
vertical tube provided a draught which sucked the combustion gases up the
tube; and this draught could be controlled so as also to control the rate of
combustion if necessary.

Architects seem to have an in-born aversion to designing a chimney,
and they tend either to ignore the need for it or design it to be unseen.
However there are many old large houses with magnificent tall chimneys
reaching well above the roof. We know that the house owners did not like
the smell in the early days of coal smoke; but that was before everyone
began to bum coal because it was much the best fuel they had ever seen. The
aristocrats could not uphold the standards of clean air which their own tall
houses violated only occasionally. Even tall industrial chimneys were
thought expensive and smaller ones were allowed to become very numerous
before sufficient political will was aroused for real prohibitive legal action to
be taken.

2.2   THE TWO AND A HALF TIMES RULE

Among the factors to be considered in building chimney stacks the 'two and a half times' rule was promulgated to advise (not to compel) architects of the need to recognise the industrial chimney as a legitimate and necessary part of the building; with a hope that with that ratio of chimney height to building height the nuisance of chimney smoke would be made tolerable at ground level in the immediate vicinity of the building. In practice most chimneys have not been made high enough on houses and office buildings because they have been thought to spoil their appearance.

2.2.1   A modern house inviting the smoke inside. A miner's new cottage in the city of Durham, England.

2.2.2   Soot in a chimney on fire filling the street with smoke, as seen from a train.

2.2.3   Smoke being carried down into the street before the introduction of smoke control.

2.2.4   In the Welsh countryside - not very neighbourly.

2.2.5   Better 19th century ideas; now preserved and 'listed' to prevent their demolition.

2.2.6   Breaking the rule at a brick works; a mixture of flagging and fumigation.

2.2.7   Oxygen lancing at Scunthorpe steel, again not tall enough for those buildings soon to pollute the ground.

2.2.8   Lime kilns making black smoke while warming up with chimneys far too short. (Buxton, England).

Tall stacks have been tolerated for industrial furnaces, and Professor Brunt lent his authority to the rule (1932), knowing that architects would assume that it had a good theoretical backing, which it actually lacked.

Chimneys can be expensive and any designer has to make some sort of compromise between the costs and benefits of very tall chimneys. The rule has been justified on the ground that it prevents the effluent from entering the turbulent wake of the building containing the furnace except in the case of the taller buildings (which office and residence blocks often are).

It was essential to develop a background of theory which fitted in with observation of model experiments, although at first the effluent was assumed to behave like a series of separate puffs each of which would be carried by the wind, and it would have its own upward velocity which would be diluted fairly quickly to become negligible.

2.3   FLAGGING

The effluent ought not to enter the wake of the chimney itself because that would bring pollution down towards the ground. This is usually avoided by having a *sufficient efflux velocity*, and it is improved by minimising the size of the chimney as an obstacle in the wind. Nevertheless some architects have attempted to decorate the tops of chimneys with elaborate brickwork which has been likely to cause flagging.

2.3.1 Flagging worsened by an
        elaborate chimney top.

2.3.2   A disc to prevent flagging at Saint
Auban aluminium smelter in SE France.

Modern attempts to reduce the external width of chimneys by prefabricating them out of sheet metal were at first frustrated by the loss of buoyancy of the effluent by heat conduction through the wall to the outside air; and this has made it necessary to have an outside cylinder with an insulating air space between it and the basic flue.

2.4   STRONG WINDS

In order to achieve an efflux velocity at the stack top which prevents flagging the diameter of the mouth can be decreased by a contraction at the top; but it requires additional fan power to increase the exit velocity in this way and it may be a costly extra requiring additional maintenance. It has to be decided what efflux velocity is required, having in mind that (as we shall see) the effluent is diluted by the wind. Flagging does not create a nuisance in high winds because of this dilution, and the expense of an increased exit velocity may not then be justified. An exit velocity of 7–10 m.sec^{-1} is usually sufficient in practice.

2.4.1   A plume stretched and diluted by a strong wind in slightly stable air. A cement works emitting dust and condensed moisture (near Sheffield). As the condensed moisture evaporates only the whitish cement dust remains.

2.4.2   Cement works plumes in turbulence due to thermal convection, near Thurrock on the North side of the Thames Estuary.

2.4.3   Warner's steel works on Teesside emitting a sulphurous plume which is becoming sinuous while thermal turbulence is increasing in the sunshine.

2.4.4 Gusty thermal turbulence brings down the plume to the ground at Ironbridge old power station. The cooling by the river has since this picture been replaced by cooling towers, taller chimneys and greater production of power.

A particular problem with the heat loss through a thin metal chimney wall was that if the fuel contains a lot of sulphur, when the furnace is first lit up from cold there could be a considerable condensation of oily sulphuric acid on the inside of the cold chimney. Then when the system became warm and the draught increased to working speed these acid droplets were carried out of the chimney top and fell to the ground close by. This caused holes to be created in car roofs and clothing, and was avoided only by keeping people and property away from the area until the acid was evaporated by a higher effluent temperature and carried away as a vapour which did not soon fall out. This problem was an unfortunate experience at the gold mine in Kalgoorlie, Western Australia, where the mineral containing the gold contained an unusually high proportion of sulphur which was being exploited as an important contribution to the fuel required for the smelting process.

2.5   STACK OSCILLATION IN THE WIND

2.5.1   A steel stack with a strake on the top section to prevent vortex street oscillation. The brick stack nearby is too heavy to be forced into oscillation.

Another problem with the steel prefabricated stacks is their lightness. This gives them a fairly high frequency and amplitude of oscillation if held fixed only by their foundations. They are liable to oscillate sideways in a strong wind, throwing off a vortex street. It was feared that the foundations would be made loose. To prevent a structural failure steel ropes were often attached to counteract the sideways forces in vortex shedding. If another similar stack stood close downwind it might be made to oscillate more violently, and cases have occurred in which one or both have been rendered unsafe. Heavy brick stacks are too rigid and massive for oscillations to build up and become dangerous or disruptive. The creation of a vortex-shedding frequency of stack oscillation can be prevented on a lighter stack by the installation of a strake which causes the shedding to occur at different times at different heights and so that no resonant build-up can happen. An alternative is to place a heavy mass at the top so as to lower the resonant frequency out of the range of the vortex shedding frequencies.

2.5.2   Two views of a counter-balance mass damper which prevents the growth of oscillations which would cause fatigue at the foundations. The mass is attached by springs and does not allow oscillations of the vortex-shedding frequency to build up.
(Pictures by *Beaumont Chimneys*).

2.6   THERMAL RISE AND EFFECTIVE STACK HEIGHT

The *effective stack height* is that at which a source, of the same pollution strength but of zero buoyancy and zero vertical efflux velocity and in the same wind structure, would produce the same pollution at ground level as the actual pollution source. We would naturally wish to raise the effective height as much as possible above the actual stack height required for other reasons.

The wind structure includes the turbulence which is seen as the agent of the dilution which spreads the pollution down towards the ground.

The effective height defined in this way is specific to the occasion. The turbulence may be *mechanically* produced by obstacles and roughnesses on the ground, or by *thermal* convection. In *mechanical turbulence* the intensity decreases upwards away from the obstacles, but in *thermal turbulence* the eddy size increases upwards because buoyancy gives upward momentum to the eddies and tends to transfer pollution upwards as it is diluted rather than downwards when the source of buoyancy and of pollution are both at or close to the ground and may even be the same source, namely the chimney of a particular furnace.

A strong source of pollution placed at the top of a large building may be feeding its effluent by flagging into the wake of the building, which causes objectionable concentrations at the ground and could give the source an effective stack height below the actual source.

The effective height of a stack may be increased by increasing the upward velocity at efflux (by narrowing the chimney diameter at the top), **but** engineers must be warned that this may be a very uneconomical choice because it requires extra fan power.

2.6.1   When five extra generators were added at the power station in Kingston, Tenn., to supply the load required for separation of uranium isotopes for the first nuclear bomb in 1943, a nozzle was fitted to the fourth old stack but only raised the plume less than the increase in height of the other new stacks in a not very strong wind.

2.6.2   The jet of this steam engine's safety valve ceased to raise it after only a short rise.

2.6.3   This architect's fancy which reduced the efflux velocity also caused flagging which blackened the chimney top and parts of the building soon after the oil furnace was first ignited. There was no significant visible smoke and so we may suppose that not seeing the flagging was regarded as sufficient reason to indulge the whim.

Although increased efflux velocity does increase the effective stack height it also increases the turbulent mixing with the surrounding air and that has the opposite effect of more rapidly decreasing the upward speed as it shares its upward momentum with an increased volume. By contrast the buoyancy of a hot effluent decreases its upward momentum more slowly. The buoyancy remains with the diluted effluent and continuously adds to its upward momentum (gBt, where t is the time since efflux and B, the buoyancy = ratio of the external to the internal density) whereas any increased efflux velocity only gives it a fixed amount of upward component of momentum, denoted by w. In practice at a typical power station w becomes less than gBt a very few seconds after emission with B equal to the (temperature excess at efflux)/(absolute temperature of the air) and g is the magnitude of the acceleration due to gravity. Thus loss of buoyancy (heat) before efflux is to be avoided, and a narrowing of the stack mouth to increase upward momentum can become expensive and yet when t is no more than a few seconds we would usually find that

$$gBt > w.$$

and by considering this requirement it can usually be seen that a narrow nozzle is only worth while if it is used to avoid flagging.

2.7   THE PROBLEM OF SEVERAL FLUES CLOSE TOGETHER

It used to be thought that where the output of a power station was increased by adding more generating sets that the chimneys should be spread out to be as far apart as possible. Then the concentration of effluent when it first reached the ground would be less than if their plumes had overlapped. The basis of this idea was the assumption that the turbulence existing in the airstream was the main agent of the dilution. That this was an error in the case of a large source had not been appreciated, and there was an inherent dislike of chimneys and the treatment of small sources had avoided the two and a half times rule because they were regarded as negligible. As objections to large sources became more severe and common, power stations were allowed to have chimneys up to 200 ft high. Fulham power station shown in the picture 2.7.1 was designed to burn pulverised coal, and although the combustion was as efficient as any at that time ash deposits were a major nuisance. It was often reported that the plume was carried for many miles before it produced a detectable concentration at the ground, other than fallout ash (grit) very close to or on the station.

But the four chimneys at Battersea Power Station (seen in Chapter 4) were placed at the four corners of the building, and the relic has been 'listed' and retained some of the sentimental attraction of that faulty design.

2.7.1   Fulham Power station in a NNW wind, and almost no thermal rise.

2.7.2   The same in a WNW wind with some thermal rise as the four chimneys fed into the same plume. Most of the effluent is seen as white dust well above the chimneys; but a downgust nevertheless partially obscures some of the building.

In those pre-war 1930s objections to power stations in cities were becoming frequent on the grounds that they were responsible for fogs and squalor and so the Central Electricity Generating Board (CEGB) were ultimately required to install flue gas washing equipment in the proposed power station at Battersea. Because there were fears that too much waste heat was being got rid of into the river which reduced the dissolved oxygen with adverse effects on the fish, they were required by Westminster City Council in the originally accepted design to supply waste heat to the new blocks of flats on the opposite side of the river. The City of London was then able to require gas washing to be installed at the new oil-burning station further down the river at Bankside because of the larger sulphur content of the oil fuel.

It has been mentioned that attention was drawn during these arguments to the fact that the Fulham plume(s) rose from the 200 ft chimneys and did not pollute the immediate urban area; which implied that 200 ft was an adequate height for satisfactory dispersal. But it was not emphasised that this dispersal occurred only when the wind was along the line of the stacks so that the plumes did not remain separate! More on that in the next section.

2.8   THE ADVANTAGE OF A MULTI-FLUE STACK

A multi-flue stack has, typically, four separate flues, probably of steel, inside a brick or concrete stack of square cross section with rounded corners. The flues emerge as four separate nozzles protruding far enough to avoid flagging in the much wider wake of the quadruple stack. Each flue serves a separate furnace/boiler/generator or whatever, and if it is wished to operate on a lower load than full, less than four units may be operated without decreasing the efflux velocity and the effectiveness of the chimney.

It will easily be seen that on emerging from a single stack operating alone that the width of the plume is very quickly diluted to about five times its width. Therefore four plumes emerging at the top of a multi-flue stack quickly mix with each other so that after a very short distance they occupy a width about the same as would be taken up by a single plume. Thus four plumes behave in the same way, after that very short distance, as if it were only one plume with four times the buoyancy.

Without the helpful effect of the increase in buoyancy, putting four flues into one would quadruple the concentration of the plume when it became diffused down to the ground, and so having a multi-flue stack is only worth while if the buoyancy greatly increases the *effective* stack height.

2.8.1   Eggborough Power Station planned with a 4-flue 200 metre stack. The plume is never seen at the ground and is very difficult to identify because of other smaller sources intervening before it becomes detectable at the surface. The plume is seen normally to rise up to the cloud base.

2.8.2   Bankside Station on full load, seen from the top of St. Paul's Cathedral. The wind is very calm on this occasion so that the plume rises vertically and is not diluted as rapidly as when bent over.

In the case of a power station it is often seen that such a plume rises up to the cloud base most of the time, even when the stack height is only 200 metres, and the cloud base is at about 800 metres. Consequently a buoyant enough plume will be carried to the top of the *mixing layer* and is well diluted before being diffused downwards.

2.9   BIFURCATION OF A BUOYANT PLUME

The buoyancy of a plume causes a vortex pair cross section to be formed with visibly clear air being drawn up the centre line of the plume so that it becomes bifurcated. The bifurcation is an indication that the plume is rising, and is easily recognisable.

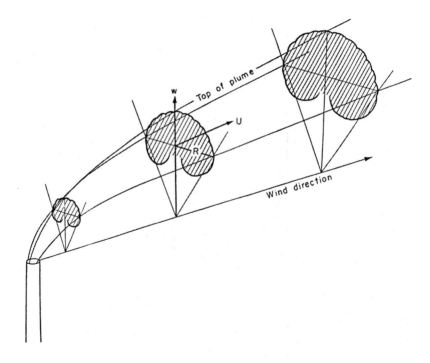

2.9.1   A perspective diagram of the shape produced by buoyancy.

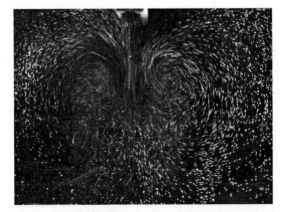

2.9.2   A short time exposure of a water tank experiment with white particles to show the flow. A horizontal line thermal of buoyant fluid was released from a trough, in a series of experiments to measure the rate of rise as a function of buoyancy (see Scorer 1997).

2.9.3   The bifurcation of two merging plumes near Scunthorpe (see also 3.3.2).

2.9.4   Two plumes at Battersea where the unwashed plume on the left rises through a weak inversion while the washed plume with less buoyancy is bifurcated as if at a ceiling.

Although there have been many attempts to find a formula for the thermal rise, and many guesses have been made to suit occasions of particular interest, it has become obvious that the magnitude of the rise depends very much on the thermal and turbulence structure of the atmosphere into which the plume is emitted. Thus, when sunshine is producing thermal convection the plume from a power station with a multi-flue stack is likely to be carried up to the condensation level (cloud base). Above that level the plume is not easily identified as distinct from small cumulus clouds. The effective stack height would normally be identified by where the base of the plume begins to be mixed downwards.

The thermal rise has been observed to be so very significant that it is often two or three times the height of the chimney itself, and this means that the pollution is often undetectable at the ground for at the very least 20 or 30 chimney heights downwind. (See sections on *The mixing layer* in Chapters 3 and 6).

We have already seen that the plume(s) at Fulham was already illustrating this feature, and the new power station at Bankside was built with a four flue single chimney the idea being to carry the plume quickly to a height above the dome of St. Paul's cathedral which is on the opposite side of the river and reaches the same height as the chimney. But a theory failed in a quite different respect, namely in allowing the wet gas washing plant to cause it to descend below the chimney top. (See Chapter 4)

3

Dispersal in theory and practice

3.1 INEVITABLE INACCURACIES

The problem has always been that a theory (i.e. a mathematical description) of how pollution is diluted into the environment must probably involve several parameters which have to be found by observation. A formula which gives a correct answer today, because the parameters were based on today's observations, is very unlikely to give as accurate an answer for tomorrow and beyond. Generally an answer is considered to be as good as can be expected if it is correct to within a factor of 2 or 3, (× or ÷); this is because any model that could be relied on to do better would have to be very complicated and have too many parameters to be conveniently measured or predicted. The parameters have to amount to a description of today's weather, and that may change in many essential ways, even during part of a single day.

An essential extra difficulty is that the scale of the dispersal we are concerned with increases with the time since the moment of emission and distance from the emission point. This means that close to the emission point we are concerned with distances comparable with the distance from the source, and not with distances and times beyond our immediate concern; but when we consider the travel of particles emitted a few hours ago we are concerned only with very rough approximations to the position. At the earlier moment we bother only about the wind direction and strength here and now; but changes of wind during the next minute, or two or perhaps further downwind, will not concern us in three hours time, or tomorrow.

Thus we are concerned mainly with eddies whose size is comparable with the distance and time already travelled. If the relevant eddy size is roughly the same as the time and distance from the source, then the sideways spread will be proportional to the distance travelled and the spread will widen according to the distance and the spread will be outward with its boundary along a cone (i.e. a straight-sided wedge, in plan). This very basic idea, first suggested by Sutton, still leaves the angle of spread as an open parameter, which must be measured by observation if at all. If the relevant time is not short the relevant angle might have to be changed if the time does not remain short; and then long term averages would have to be used.

3.2   MODELS AND LEGISLATION

In most places air pollution is a significant problem for only part of the time. Legislation (enforced regulation) then takes the form:- The air pollution concentration should not be allowed to exceed a particular specified concentration for more than such-and-such a number of hours on more than a certain number of occasions in any calendar year. If these bounds are transgressed a special kind of action must be taken, such as reducing the sources or compelling them not to emit in certain weather types which seem likely to cause a rise in pollution concentration.

Such action has to be taken because the restrictions were too strict for the case. But whatever the circumstances this book is not intended to help the authorities to decide how to deal with situations arising from an earlier decision to make a restrictive law in the first place. It may not be appropriate to attempt to harmonise the regulations for several different places which have differences in climate even though they are not very far apart geographically.

Our subject is dominated by variations in the weather and climate from place to place and time to time.

Furthermore the observations that have to be envisaged in planning any regulatory system are already installed in very few places. Models for air pollution are essentially built around a few sets of observations, and there are no generalising laws to which new observations must in all cases be fitted, yet this is the case with the production of a weather forecast on which ANY prediction of air pollution has to be based. Put very simply: the weather forecasts have origins which make them unreliable after a short time whose length would be a matter of dispute while the correctness of the observations is not expected to be disputed. Any prediction of air pollution concentration has to take the weather forecast as given, and must base another expectation upon it by means of a formula which does not have any pretence of reliability superior, or even equal to the model of the weather forecast.

It therefore becomes a matter of dispute to assert, in the circumstance just described, that the best way forward is to lay out the mechanisms which cause severe, extreme, or otherwise seriously inconvenient conditions of air pollution, and by experience learn to give suitable warnings for others to take note of in advance of their happening.

3.2.1 and 3.2.2   These two pictures were taken on the same day at 15.00 and 19.45 hrs. from a train on the outward and homeward journeys. In the first some of the chimneys of the brickworks can be seen and clouds of pollution lie between them and the train. On the return, although the plumes could still be seen to cross the track, the works could no longer be located by smell because the plumes were all rising and passing overhead with fresh air passing beneath them. The sun was setting and so there was no *thermal turbulence* being generated as the ground cooled.

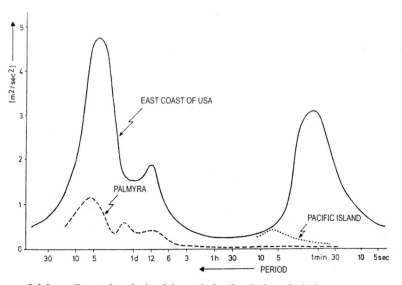

3.2.3   Spectral analysis of the periods of variation of winds.
To be suitable for treatment by a mixing-length type of theory there
should be wind systems of long duration which correspond with the mean
wind – the length of the sample being only a small fraction of the typical
variation period. Also, there should be periods of much smaller length of
which a sample would measure an average of several periods, so that the
domination of the sample by a part of one short period or even a small
group of short period variations cannot dominate the sample.
In this sample the main curve represents Brookhaven, on the East coast of
U.S. where the dominant periods are about 5 days, and 2 minutes. But the
amplitude does not fall to zero in between, so that a 1 hr sample would
not represent anything typical and could be very misleading.
The dotted line represents the spectrum for a Pacific island where tropical
rainstorms are dominant in the doldrums.

It is, of course well known that the severity of pollution phenomena
is one group of meteorological events which is subject to rapid diurnal
variations which require major disruptions and alterations to the
mathematical models to make them applicable in the change of
circumstances with time of day.

An important body of theory has developed under the heading of
Similarity theory and this is described by Scorer (1997) under the heading of
Partly turbulent flow and it gives many power laws which it is tempting to
use when the temperature and velocity profiles do not fit the simple profile
laws strictly required. Another well known example of the extension of a
theory applied usually outside its proper circumstances such as when the
Richardson Number of the airstream profile varies with height and is
effectively assumed constant. It is dubious in use because of assumptions

about the conditions at the boundaries. Likewise, when a stably stratified layer generates a field of *Billows* resort has to be made to crude descriptions of the nature of the flow as it becomes turbulent, and almost the same goes for the 'breakdown' of a viscous boundary layer, and one in which a horizontal stream of liquid flows against a strong wind and the flow is made to separate from the free surface producing temporary smooth or disturbed parts of the surface.

We now revert to one of the simplest air pollution source models!

3.3   DISPERSAL OF A SOLITARY PLUME: THEORY

If the turbulence were such that it produced results similar to a single constant diffusion coefficient, the width of the plume would be proportional to the square root of the distance from the source. The following is an argument for Sutton's theory that a diffusion coefficient should be linearly proportional to distance (or possibly to time) from the source. As a deterrent to too much faith in theories it should be noted that in spite of the long theorising at Porton Down the British army was forced to advance through its own poison gas cloud when it was first used in battle. The direction and strength of the very light wind specially chosen for the occasion had been incorrectly forecast.

In practice, to use a diffusion coefficient it would require the eddies to be small and numerous so that any sample measurement would include a large number of eddies and would not be distorted by taking in only a fraction of one of a small number of larger eddies. This idea is usually expressed by saying **that in the spectrum of eddy sizes there must be an eddy size with no eddies close to that size.** A sample would then take a large number of those of smaller size and only part of one of a larger size. That larger one would represent the (mean) wind at the moment and place of the observation. The smaller eddies would be represented by a spreading coefficient which is assumed to produce results analogous to molecular diffusion or heat conduction. (There is a good body of mathematical theory for such cases.) The pollution is assumed to be carried along by, and be part of, the air motion. This 'mixing length theory' has to assume that that type of thinking, with a gap in the spectrum, can be applied. Indeed it would be required that a gap is always there, increasing in size with distance from the source – a very unlikely circumstance!

It is closer to reality to think of a plume as an irregularly sinuous plume of pollution which is represented by an average of the many positions of the plume over the time taken for several sinuosities to pass the point of observation (or prediction).

Thus we impose on our model the concept that the actual transport of the pollution may be described as composed of *the mean wind* with a system of *random fluctuations*, which we call turbulence, added. There must exist a spectrum of variously sized 'eddies' such that this turbulence causes many fluctuations in any sample and a *steady mean wind* is added. Or we should assume that the dilution takes the form of classical diffusion with a coefficient proportional to distance from the source.

As no justification for that proportionality had been offered, in writing what became a 'classical' text-book Pasquil (1962 and 1974) simply plunged in with the advice to use the standard diffusion equations with a constant coefficient and a Gaussian profile. The only guidance on the size of the coefficient was that it was likely to be much larger than the molecular value. It was based on the analysis by Reynolds of the stresses produced by the theoretical fluctuations, and there was a tendency to suppose that the transfer coefficient would be the same for every 'transferable' quantity, because it was thought that such was indicated by the classical mixing length theories. Some discussions assumed that the diffusion took place as if there was a diffusion coefficient for the atmosphere which only had to be measured!

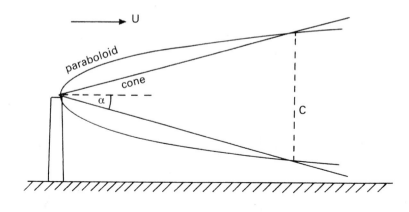

3.3.1   A paraboloid would be the shape of a plume dispersed by a constant coefficient of diffusion with its width increasing like x (x = distance downwind from the source). We compare this with a conical boundary with constant angle α which encloses all the pollution emitted. Accepting that this is more like what we see, we would say that this represents a conventional diffusion coefficient increasing linearly with distance.

3.3.2   A typical plume showing a conical shape, but there is some thermal rise of which we shall take note. *Keadby Power station* near Scunthorpe. See 2.9.3.

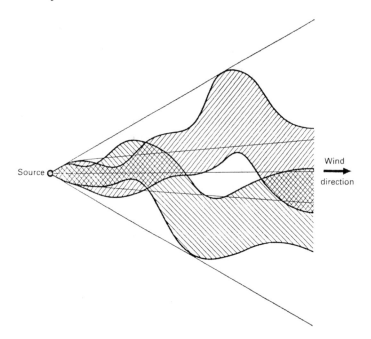

3.3.3   Plumes are often sinuous but can be represented as passing along a cone on the time average. At first the plume is dominated by its efflux with very small eddies but it may soon take up a widening flow made sinuous by the mean wind.

3.3.4   This colliery was burning its fuel inefficiently because it did not have to pay for it; the air supply to the boiler was so that some of the carbon dioxide was reduced to black smoke! Note that the material of the plume does not travel along the sinuosities but that the sinuosities travel along with the mean wind.

Our theory will be required to provide a formula for the concentration of pollution across the section of the plume at C (in 3.3.1); and considering the essential crudity of the model we may choose any suitable formula for the profile.

Thus the position of the plume is represented by an average in time, which is also a fair representation of its nuisance value for planning purposes. The conical or wedge shape is, of course, an average or mean position imagined instantaneously in the existing wind structure which may have become quite different by tomorrow (or even in an hour's time!).

Every wind structure has variations and a single profile has to represent its average over an appropriate stretch of time. The concentration may frequently be zero at places within the wedge and may have a maximum, which only lasts for part of one sinuosity but might sometimes rise to 4 times the value of our average at a particular position. The effect on different people may also be significant. It is a common experience for a smelly pollution source to be imperceptible if it remains at a constant strength for people in the plume, but may become a source of complaint if the average remains the same but the strength varies by a factor of 0 to 10 in the short term. Some oil refineries have been said to smell like badly burnt onions, and it is interesting to note that the complaints come mainly from the edges of the spreading plume where its incidence is intermittent. Can we trust our noses? An occasionally strong acid plume may be bad for sensitive throats. A generally objectionable odour may nevertheless be acceptable to some people (especially if it is associated with their source of income).

There used to be a factory making vanilla concentrate in south Wimbledon, of which the smell was perceptible out of doors at a distance of over two miles downwind: although when an employee called at my house (at about 2 miles) he said he could not smell it anywhere outside the factory, yet I could smell it from his clothes. He told me that when he went on holiday he became more sensitive and could detect it heavily in his wardrobe when he returned home.

From this it is clear that an unfamiliar odour is very readily noticed and that the frequency and severity of complaints may be of uncertain meaning in terms of the real incidence.

Any measurement of the parameter that is being forecast in any formula that has a statistical basis must be made so as to contain an appropriate sample. We are arguing that the phenomenon is being caused by fluctuating air velocities and the appropriate sampling size grows with time and distance from the source. If observations are being used to make a test or validation of the theory or formula they MUST satisfy this condition: the sample size is proportional to distance or time from the position and moment of release at the source. This is a requirement for models used for forward (or backward) tracking of pollution particles assumed to be travelling with the air in which they are carried but with a turbulent motion which does not affect the tracking of the air whose particles have a different sort of identity which is not carried by the turbulence.

3.4   MAXIMUM GROUND LEVEL CONCENTRATION (MAX GLC)

Although this may not be a very accurate measure of the strongest whiff of a plume that may be experienced in the neighbourhood, it could be a useful measure for comparison between different tall chimneys in planning.

If we assume that the plume is conical with apex at the *effective chimney height,* and has a horizontal axis, then at the point of first arrival at point A at the ground its vertical cross-section has a radius equal to the effective chimney height, H. The vertical section of the plume has area proportional to H^2, the constant of proportionality depending on the shape of the cross-section. With an average wind speed U over the cross-section and a suitable average concentration of pollution P, the flux Q, of pollution along the plume must be

$$Q = PH^2 U$$

Observations of real plumes make us realise that for any one occasion the approximations we have to make to get any formula which contains a useful lesson for us must be very simplistic. Otherwise the weather forecast for the occasion would have to be automatically represented in making the choice of parameters.

It is true that the cone- or wedge-shaped model is sometimes quite valid when there is a wind stretching the plume out horizontally. Thus there is no pollution near the stack base until a distance is reached where the cone makes its first intersection with the ground. From this point down wind an assumption has to be made about the proportion of the pollutant which is absorbed on the ground. The part that is not absorbed is presumably carried upwards by the same diffusion mechanism as that which brought it down. It is convenient therefore, to imagine a second source in the position of the image in the ground of the actual chimney, and to add its cone-full to the original cone-full. In the case of no absorption, at the ground this simply means a doubling of the value (at the ground) from where the plume and its image first meet.

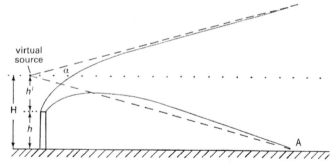

3.4.1   The plume entering its horizontal cone or wedge of angle α, ultimately to impact on the ground at point A. The apex of the cone is the virtual source, and is located at the effective height H.

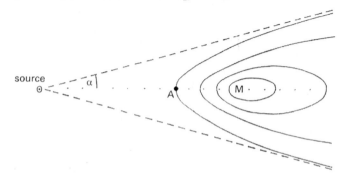

3.4.2   The *footprint* of the plume, with pollution contours increasing from the point of first impact A, to the Max GLC at M.

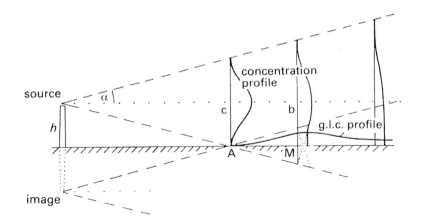

3.4.3   The vertical cross section of figure 3.4.2 with the image source in the ground and the probable shape of the concentration profiles at cross-sections c and b and on the ground.

How do we take the model any further, which we may well wish to do, particularly if we are *back-tracking* to find the probable source of pollution we have actually measured, or if we are forward tracking to find to where the pollution from a known source is likely to be carried, and whether it easily explains a recent observation?

It is necessary to make a guess at the distribution of the pollution in a vertical cross-section above the position M, in figure 3.4.3, in order to estimate the concentration at M. In practice this requires a choice of formula to describe this profile. A popular choice is the Gaussian distribution because it goes well with a diffusion coefficient and that is simpler for people who have worked in conduction and diffusion problems. An array of possible profiles is illustrated in figure 3.4.4, and each is provided with a formula which can only represent an average of values likely to be experienced as the sinuous plume fluctuates in its travel downwind. The obvious next step is to imagine that the sinuous plume also has a Gaussian profile and work out what its random twistings would add up to. The crosses on figure 3.4.4 represent what could be a collection of observations which would be required to fit the formula if it were strictly correct; but they are not usually made simultaneously, although they would have to be to represent an instantaneous profile, with a sampling time proportional to the time travelled since release at the source.

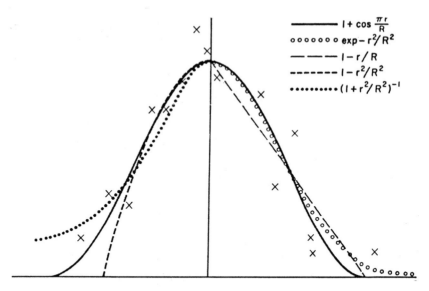

3.4.4   Examples of possible mathematically defined concentration profiles with a mathematical shape which would have to be assumed to apply for the whole length of the plume as a very crude average. r is co-ordinate distance from the middle and R defines the distance from the middle of the edge of the plume (the boundary of the cone).

3.5   THE MIXING LAYER

3.5.1   The plume from this cement works near Sheffield has enough buoyancy to overshoot into the potentially warmer air above before sinking back with the haze top as ceiling of the mixed layer below.

A much more important topic than the theoretical outline of the conical plume model is the *Mixing layer*. It originates from the frequently made observation of an inversion which also appears to be the top of a layer into which the pollution is well mixed but above which there appears to be none or almost no pollution.

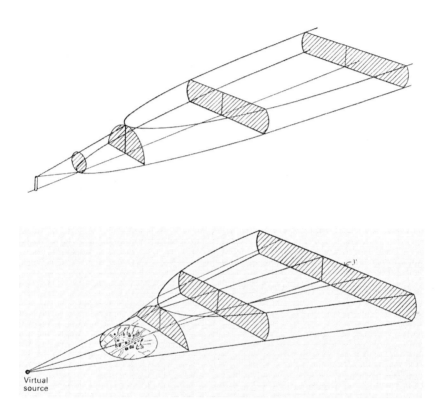

Virtual
source

3.5.2   Under a low inversion top the cross-section becomes a wide strip with its area proportional to the width y. Thus the area increases linearly when it grows only sideways and not upwards from a mixing layer of uniform depth. If the source is an area source like a town it will have a virtual source some distance upwind, as shown in the lower diagram.

3.5.3   The mixing layer seen at the South coast of England with large
          cumulus developing above the condensation level.
          (*P. M. Saunders*)

3.6   THE SUB-CLOUD INVERSION

3.6.1   Well defined haze top in the valleys of western Europe.

By the profuse supply of visible pollution in the form of cement dust and some condensed water vapour in the white plume we notice that the haze top appears to be at about the level of the surrounding hill tops.

Quite often there appears to be an inversion at the condensation level. When there is a development of smallish cumulus there appears to be a sharp haze top at that level. We call this the *sub-cloud inversion* and it is an important concept because there is a mechanism which, like the high points of surrounding hills, seems to stabilise the top of the air below that level.

3.6.1.a   A close-up of a bit of 3.6.1.

3.6.2   A *subsidence inversion* over a large part of Southern England in an anticyclone.

3.6.3   Sub-cloud inversion forming a defined mixing layer, with clean air above Ironbridge New Power Station.

3.6.4   Tower of smoke that penetrated the sub-cloud inversion because it originated in a large forest fire in Colorado, and the thermal cumulus mixed the smoke into the air above and it did not descend again.

3.6.5   Shadows of tiny cumulus on the pollution in the sub-cloud layer, and the sharpness of the more distant outlines of their tops make a contrast with the air below the inversion over Heathrow airport.

There are other causes of the occurrence of haze tops such as *subsidence inversions*. Sometimes the formation of a mixing layer can be seen in the shadows cast on the pollution by the cumulus fragments which are the first to form in the morning.

3.7   NON-DISPERSING PLUMES ALOFT

3.7.1   One of the world's tallest chimneys, whose plume slightly overshoots an inversion before falling back to a steady level at the metallurgical mine at Sudbury, Ontario. *A. & J. Verkaik*

Some of the most spectacular chimney plumes are trapped under inversions to become the first sign that there is a stable layer there. The model in which it is assumed that all plumes continue to spread to wider proportions becomes obviously inapplicable.

3.7.2   The power station plume at Fairbanks, Alaska, in mid winter where a car exhaust smog persists under a cloud made by the plume. A local problem is 'how to go shopping' – Stop your car engine so that it freezes solid, or leave it running and aggravate the smog with exhaust. *CE. Benson.*

3.7.3   The plume one early morning in the Severn valley at Ironbridge. This old station used the river water for cooling, and caused steaming fog on the river. More lately a new Power Station has cooling towers which put out visible steam at a much higher level. (See picture 2.4.4)

3.8   NON-DISPERSING PLUMES AT THE SURFACE

 3.8.1 The paper pulp works plume at Fort Townsend on Puget Sound is carried across the cool water to Seattle. The plume quickly ceases to widen until it reaches the far shore, where the rough ground causes it to begin to spread as if it had only come from next door. It has lain on the water all the way with no dilution by mixing.

 3.8.2 When the oil tanker Torrey Canyon was wrecked on islands off the coast of Cornwall the oil began to foul the local coasts. The government ordered that it should be bombed with incendiaries so that the escaped oil would be burned off. They had meteorological advice that the polluting smoke would be carried upwards by the heat of combustion and dispersed before it could do damage. To test this advice cold oil was ignited in a large tank and was allowed to burn for several hours in southern England and the smoke, seen in this picture, was not detected anywhere at all at the ground within 30 miles.

This situation is common when pollution is released at a low level and is spread into the whole depth of the mixing layer by thermal turbulence during most of the day, but as the solar heating is reduced in the evening this form of turbulence ceases and the pollution hangs around on the ground. If the emission from tall chimneys reaches the top of the mixing layer at this time it forms a polluted layer which will cause *fumigation* after sunrise the next morning, probably in quite a different location several miles away.

3.9   THEORETICAL SECOND MAXIMUM AT THE GROUND

If the lid of a mixing layer is very sharp it may prevent any of the pollution from being mixed into the air above, so that the turbulent motion which mixed upwards will equally mix it downwards, and provision for this process may take the form of an upside down image source at the inversion.

The occasion described below was discussed at a meeting, but it was considered possible that the variations which are normally to be expected could well be the actual event which happened. It will be seen that there have to be many phenomena which should be included in any model and for this reason it has become customary to be satisfied with observations which are within a factor of 2 or 3 of the theory.

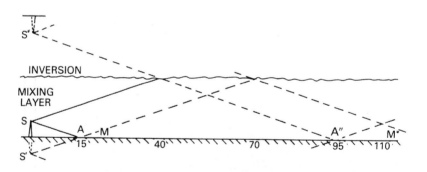

3.9.1   The figure is drawn with chimney height so as to indicate the expected scale of the effect. The horizontal distances are in chimney heights. There appeared to have been a case of this recorded on a night when the plumes from the Thames side power stations were being sampled and surveyed when there was a strong inversion. The first impact on the ground is indicated at A at 15 chimney heights and the first maximum at M, so that the second first impact and second maximum indicated at A" and M" at around 100 chimney heights distance. *D. Moore APER, CERL*

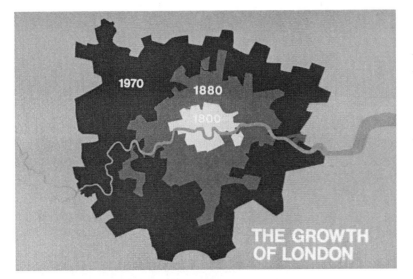

3.9.2 The growth of the administrative area of London creates new problems in defining an area source. The weather parameters may not be the same over the whole area.

3.10 COMMENT: THE THEORETICAL BENT-OVER PLUME

In general it does not appear useful to depend on a model as a forecasting tool. If a meteorological situation can be identified in which pollution problems will appear a model may help to compare one plan with another for these specified weather conditions to define features to be built into a project. Wind tunnels may help in designing chimney outlets or air intakes on large buildings.

Models may be used for comparison of two sites whose weather records have been compared and found to be very similar. A recent additional development at one may be used to predict probable effects at the other.

A feature of plumes is that they are sometimes the only part of the local atmosphere that is turbulent. The environment varies in the pattern of its motion seasonally, diurnally, hourly, by the minute, and even more frequently than that, but is only explored by suitable instruments and ultimately to the scale of molecular diffusion. Sometimes we experience spells of weather for a few days; but there certainly exists no pattern predictable reliably which can be measured up and used as a regular stereotype (except for molecular diffusion as a function of temperature) There is therefore a challenge to make a collection of 'cases' which can be often found occurring and we now illustrate how the bending over process can be broken down into a sequence of cases. We have to choose from the various cases of *partly turbulent flow* namely:- turbulent jet, turbulent vertical buoyant jet (plume), each with point source, or 2D line source with cross flow; also the instantaneous point and line 2D sources. Each of these produces its characteristic motion pattern which has its own way of evolving and retaining similarity in uniform and otherwise motionless original state. They are described in detail in Scorer (1997).

Of these we use the vertical jet (picture 2.6.2) and the 2D line thermal as used in diagram 2.9.1. In order to use the model we assume that if the ambient turbulence has a typical velocity this will not disturb the plume until it has been diluted so that its own characteristic velocity has become comparable with, and is then swamped by, the ambient turbulence.

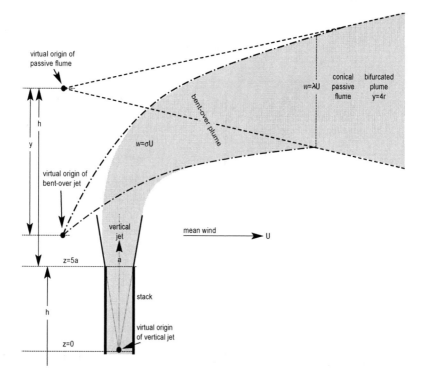

3.10.1   Referring to the diagram, it is noted that the chimney gases emerge as a vertical jet which has a virtual origin (source) at a depth $5a$ inside the stack, widening at a constant rate so that it drifts downwind as it is diluted with ambient air. At the same time its upward velocity w becomes a smaller multiple σU of U, the ambient wind. As soon as σ becomes less than unity the plume is rising more slowly than the wind is carrying it along and so it is tilted over at more than 45°. When the dilution becomes such that its rate of rise is less than the typical ambient eddy velocity it will be diffused by the ambient turbulence and so according to our modelling it will be diffused into its bounding cone.

This part of the model is in conflict with the idea that buoyant plumes inevitably become bifurcated and have a cross-section growing as is shown in diagram 2.9.1. and retaining its characteristic shape; and in this part of the plume we are assuming that it grows as if it were a section of a 2D line thermal. i.e. as if the cross-sections were the same size as it, both upstream and downstream instead of actually growing along the plume.

We shall finish this chapter with a survey of the changes a plume goes through in the transformation to being bent over. The following are the stages:

It begins as a vertical jet which widens as it mixes with the surrounding air. If the cross wind is very light it may be a few seconds before most of the plume has taken up the horizontal velocity of the wind after which we can describe it as *bent over* when it is at more than 45° to the chimney direction. The vertical velocity of the plume as a whole becomes less than the wind speed because of mixing and dilution. The buoyancy-made upward velocity may still be more than the eddy velocity defined by the turbulence coefficient. But when the turbulent velocity becomes greater than the upward speed of the plume as a whole it will begin to expand downwards by mixing, and now is the moment to determine the effective height of the chimney. In order to carry out this analysis many factors must be known in advance, and they are not trivial. (An abbreviated version of these requirements is given in Scorer (1997).)

Tall chimneys have a great advantage in being above most mechanically produced turbulence and in being above thermal turbulence for a significant part of the day and usually all the night.

Thus the virtual origin of the plume moves from inside the stack to the source of the bent over jet and then to a position at its effective height.

Also we have used the idea that the turbulent velocities in the air are some knowable multiple of the wind speed, U, which would only be true if the turbulence were of mechanical origin and there was no thermally generated turbulence such as would produce sinuosities in the plume like those of picture 3.3.4.

Only one of the plumes in picture 2.6.1 is behaving in the manner of this model. The calculations show that the vertical jet and the bending over process add very little in practice while a thermal rise may often contribute 2 or 3 chimney heights which are important because the plume will not be detected at the ground until a distance of 20 or 30 chimney heights, if then. (See Chapter 2 section 7) .

3.11 THE NATURE OF ACTUAL TURBULENCE

When a plume becomes '*bent over*' it is handing over its pattern of behaviour from the conditions that pertain at emission to the passive condition in which it is carried by, and with, the ambient air.

The turbulence is superposed on the presumably smooth horizontal wind. It is not suggested that every difficulty can and must be catered for if a model is to be acceptable; but a severe critique of a model must be carried out before any advice is based on its calculations.

Turbulence is of two origins

(1)   the *mechanical* disturbance due to flowing over '*rough ground*' which includes buildings, trees, cliffs etc. The turbulence is the wakes of these obstacles; and the turbulence 'begins', or seems to originate where the flow first separates from a rigid surface, or perhaps on a free surface of discontinuity of density or velocity.

Mechanical turbulence decreases in strength away from the place where it began or first appeared by separation. It is subject to the laws of spontaneous decay which are properties of homogeneous isotropic turbulence. That means that its kinetic energy dies away by transfer continuously to smaller eddy sizes, ultimately to become heat through viscosity fairly quickly, and must have a continuing cause if it is to be kept going. It will in general disperse pollution into a greater volume irreversibly both in the sense of diffusion and in the sense that a wake widens.

(2)   the disturbance of *thermal convection* which depends on the time of day or anything else which contributes to the determination of the surface temperature.

Thermal turbulence is continuously generated where buoyancy is present but not uniformly so. Normally a chaotic buoyancy distribution produces chaotic vorticity. However a slight departure of a buoyancy force from equilibrium may produce no more disturbance than the transmission of stable waves.

When thermal convection is generated with thermals rising upwards the individual (isolated) thermals grow in size upwards, although the upward velocity decreases. The generation of vorticity ceases if the buoyant fluid reaches an equilibrium with surroundings of the same density. If it arrives at the level of equilibrium with a significant upward velocity gravity waves will be generated which will travel away horizontally.

Bodies of heavier fluid move downwards with similar velocity patterns, with wet flue gas washing being a very relevant example.

From the point of view of aviation the most serious downdraughts are produced when shearflow causes a rainstorm cloud to lean over and allow the rain to fall through outside drier air and cool it by evaporation to produce a downburst. When the downburst arrives at the ground it may cause very sharp-edged horizontal gusts which are a particular danger to aircraft attempting to land. These downdraughts may signal the arrival of cleaner air to replace polluted air.

Turbulence is often characterised by a coefficient which is used to describe the magnitude of the velocity fluctuations as a proportion of the wind speed; and this is more appropriate for mechanical turbulence than for thermal turbulence. Or is it a useful concept applied to turbulence in the free air away from rigid boundaries, at high altitude; where the aircraft air speed is a most significant factor?

4

Acid rain and development of gas washing

4.1   ACID RAIN, AND THE HEALTH OF FRESH WATER FISH

It is inevitable that most, if not all, rain is acid because of the CO_2 in the atmosphere. Although there is uncertainty about the amount of carbon dioxide that has been present in distant ages of the past, this acidity is important today because it releases the (so-called mineral) elements present in the rocks and facilitates the formation of soil. Small quantities of trace metals are spread throughout the biosphere so that the chemistry of the soil has become important in the growth of all vegetable and animal life.

The use of wood and fossil fuel of all kinds by people restores to the atmosphere large quantities of water vapour and carbon dioxide because the fuel is largely composed of hydrogen and carbon. Some oxides of nitrogen, collectively denoted by the formula NO_x, are also introduced by the use of air – which is mostly nitrogen – to supply the oxygen for combustion. There is also a continuous emission of NO_x by the degradation of vegetable refuse on the land, and some of this becomes nitrate which is dissolved by rain and carried into rivers, and from there into many urban water supplies. Nitrate (very much unwanted) also appears in well water which is replenished by drainage through agricultural land. This is a nuisance aggravated by the excessive use of nitrogenous fertilisers on farmland, and it concerns us because it also increases the emissions of NO_x to the atmosphere.

These emissions can be important when they are carried up into the stratosphere, where they may modify the equilibrium status of the ozone. But much more important is the oxidation of nitrogen in internal combustion engines in aeroplanes and cars and emitted in their exhausts. As it happens this worry is not serious because most is removed from the troposphere in rain and most planes fly in the troposphere. Thus there was considerable discussion at a time when there were plans for a very large number of commercial supersonic planes flying in the stratosphere: but the number has remained very small mostly for economic reasons.

Although the emission of CO_2 is important in the context of climate change the acidity of rain has been greatly increased by the emission of NO_x

which causes NO_2 in particular to be dissolved in cloud droplets, and consequently in rain. The other oxide of importance is SO_2 which is the result of burning large quantities of coal and oil which contain very variable proportions of sulphur. This gas has been used widely as a tracer for the detection of the plumes from power stations in Britain and Continental Europe. A major obstacle to using SO_2 as an indicator for the measurement of the dilution of a power station plume has been that it is reactive with and is absorbed by many different kinds of vegetation, including grass and common hedges. As a consequence of this and the apparent absence of significant damage to crops it was a surprise that people in Scandinavia began to complain that acid rain was killing the fish in their rivers.

There is considerable difficulty in demonstrating that the fish kills were due to the acidity of rain, furthermore it is very difficult to show that the acidity was due to acid emissions from particular places. The composition of river water in Scandinavia was measured typically only once per week. The rain occurred on particular days which did not seem to be followed immediately by the fish kills. The rivers of industrial Europe and Britain, where the concentration of acid pollution would have to be greater than in Scandinavia which was much further from the sources of the acid gases, did not experience fish kills. Also the concentration of acids in the air over Scandinavia was almost always less than was found than over the alleged sources. Indeed the acidity in the rivers did not seem to be related to the observed acidity in the air (mostly at ground level). The fish kills had occurred early in the twentieth century from time to time, but now some lakes were becoming more acid and the breeding sites for some fish were failing to produce new stocks.

The British CEGB (Central Electricity Generating Board) began research at a small mountain lake, Loch Fleet, in SW Scotland, where the inflowing and outflowing streams were monitored in detail. This was after it had become known that the fish kills were almost exclusively of trout and that the total of fish taken from the rivers in Norway had actually increased during the 1940s and 1950s while the argument had continued. Indeed it became evident that the argument was about the fly-fishing sport which had flourished previously and was not being supported because of the killing of trout species.

A similar situation was developing at the same time in the mountain streams of New England and other regions of the north-eastern corner of the American continent mainly as a result of the acid effluent from the industrial development crowded around the shores of Lake Erie. That lake tended to be picked upon because the water was being much more directly polluted by liquid industrial effluents. The rivers and lakes of the Adirondack mountains which lie in the northern half of New York State were regarded as exemplifying the areas most seriously polluted by acid rain. The densest industries and most thickly populated areas of the United States and Canada lie between South and West from these mountains, and this is the quarter from which the wind most commonly blows.

The area enclosed between Chicago, Cincinnati, Washington, Boston, and Toronto frequently has a mixing layer which is very highly polluted so that air travel does not offer the delight of seeing the home of many millions of people from the air. The same is true of much of the densely populated parts of Europe. The proportion of sulphur in the coal is highest in the areas where brown coal (lignite) is most used for electricity generation and where bituminous coal is used in the older industrial plants where the most modern processes are not yet in general use and where the public is not yet convinced that the expense of preventing air pollution is regarded as economic simply because it has been tolerated and looked upon as a necessary consequence of carrying on the industry as it has evolved. The classic example is London where there was a disastrous episode in the first week of December in 1952 which prompted political will to get rid of domestic smoke in smoke control areas through the Clean Air Act of 1956. Although the visibility has been enormously increased and the effects of making the towns very dirty has been largely ended, there still remain the effects of several gases which increase the acidity in some places that are not quite expected. Furthermore that there would be radiational effects of global warming had been foreseen for a long time but is now threatening to become serious, and details are being revealed.

Complaints about the acidity of rivers and lakes becoming acid to the detriment of fish in the NE of North America and in Scandinavia led to the accusation that the industrial areas 'upwind' (i.e. to the WSW of the complainants) were responsible and led to extensive measurements and other observations of the effluents of the larger industries. Attempts to follow plumes across the North Sea and to correlate the acidity in rivers with the dropping of rain have not found any success.

There were many difficulties among which was the problem of identifying occasions on which the rain fell with an associated fish kill not much later. Mostly there did not exist a full enough range of measurements, with river samples never being analysed more frequently than once a week and no fall, but rather an increase, in the total of fish caught for food. The acidity of rain was usually less than that of the ground or water which it was supposed to be acidifying. The 'prevailing wind' was not readily traced back to the sources which were supposed to be the origin of whatever was a major problem. Studies of wind roses show the frequency of arrival direction and strength only; and do not show the correlation of strength and direction with what the wind brings in the form of pollution nor its more distant origin. If the direction is divided into eight compass directions, each with a 45° span, the octant of greatest frequency may occupy much less than a quarter of the time. It may be argued therefore that the two adjacent octants S and SE would bring much more acidity from the industries of SE Europe into Scandinavia than the octants W and SW although the latter may have a greater frequency of occurrence. Furthermore the winds from the SE are usually much more stable and have a much shallower mixing layer than winds from the SW. If to these uncertain and unknown correlations are added the much greater frequency of observations of the weather than of the variations of acidity in the rivers and lakes the argument has not really progressed at all through the medium of routinely collected statistics.

It was found, however, that the ground in many parts of Scandinavia was far more acid to a greater depth than the ground in the industrial areas from where the acidity was being imported. This was due to the absence of calcium compounds in the soil, and by dropping rocks of calcium compounds into the acid lakes it was found in Sweden that the pH of the lake water could be raised.

4.2   THE LOCH FLEET STUDY

The CEGB conducted a long survey of Loch Fleet a small mountain lake in SW Scotland which showed that trout was the main victim and that it could be saved by putting limestone in various forms such as dust and small rocks into the tributary streams according to the requirements of the season. The Loch Fleet study is a good introduction to the whole subject of the welfare of the fish, the complaints and remedies; and it does not suffer from some of the inadequacies already described. On several occasions it was noticed that if a block of limestone was placed in a stream the trout would come and stay for long periods in the downstream wake of the stone and that the breeding took place some distance up the small streams which flowed into the lake. If the whole basin which collected rain flowing into the lake was sprayed with limestone powder and some trout was put in the lake the breeding was soon flourishing. Thus the protection of the breeding streams from acid gave the most important benefit.

4.3   THE CONCENTRATION OF ACIDITY

As usual the answer to any air pollution problem is to reduce the emissions. In this case it would have been necessary to be certain where the sources of acidity were to be found; and anyway acid gases of sufficient strength to produce the observed results had not been observed crossing into the polluted territory. It was argued that in winter particularly winds from the Eastern half of Europe which often blew in very stably stratified air increased the deposition of acid droplets from clouds onto vegetation in Scandinavia.

Snow may be deposited during the coldest season on the higher ground in winds from almost any direction. If the snow is formed in either convective or in frontal clouds it will contain many salts from sea spray and dissolved acid gases as well as acidic smoke from densely populated or industrial regions when the droplets are still at a low altitude. When the droplets freeze the salts are separated out of the ice crystals in the snow that is ultimately deposited. When the first meltwater runs out of the snow a large fraction (perhaps as much as three quarters) of the soluble acid pollution is carried into the streams. This is enough to cause a fish kill; and the origin of the acid which kills the trout may be quite unrelated to many separately detected periods of acid rain. Fish kills have become more frequent since the increases of population and industrialisation in the twentieth century.

4.4   THE BATTERSEA SAGA

It was considered necessary in the 1930s to build new power stations in the built-up areas of London where a demand for electricity was growing rapidly. The River Thames would provide cooling water and the Battersea station would have 4 generating sets which can be seen in the visible outline each to have a chimney on the top of a large wide chamber.

Putting the four stacks at the corners of the main building placed them as far apart as possible so that the dilution of their plumes would be maximised before they became additive, thus reducing the maximum concentrations to be encountered at the ground. It will be remembered that the Fulham Power Station on the opposite (north) side of the river had four chimney stacks in a row and not widely separated and with a height of around 200 ft and no concession, at last, to architect's prejudices against making chimneys visible on the skyline. A greater height might have been regarded as a danger to aircraft on the approach to landing at Heathrow Airport. (See pictures 2.7.1 and 2.7.2).

At Battersea the flue gases would enter one of the chambers at the base and rise up through a downpour of water sprayed from above, and this would dissolve out the SO_2 and carry it into the river. That part of the process was fairly successful and removed typically 96% of the SO_2.

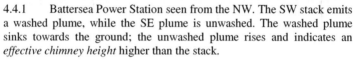

4.4.1   Battersea Power Station seen from the NW. The SW stack emits a washed plume, while the SE plume is unwashed. The washed plume sinks towards the ground; the unwashed plume rises and indicates an *effective chimney height* higher than the stack.

It is also necessary to examine the heat transfers which determine the plume temperature. It is a theorem of latent heat that the air cannot be cooled below its wet-bulb saturation temperature by evaporating water into it. This is indeed the basis of the wet-bulb measurement. It was expected, therefore, that the flue gas would emerge from the chimney top at its wetbulb temperature. In the event it emerged as an unusually bright white cloud, and this was proclaimed publicly as demonstrating that it had been washed very clean – by contrast with, for example, the plume from the Lot's Road station nearby which powered the London Underground trains and often emitted dark smoke for a few minutes of soot-blowing. However that station has since been converted to burn natural gas and is smokeless.

The Battersea plumes contained a cloud of very small droplets which were produced like any natural cloud – by the gas being cooled. The bright white cloud due to scattering of daylight was enhanced by the enormous number of condensation nuclei to be found in the flue gas from any combustion process. When this water cloud began to evaporate as the plume mixed with the unsaturated ambient air it was cooled to a temperature colder than the outside air and immediately began to sink towards the ground.

4.4.2   Two pictures of the Battersea plume seen from the opposite
bank descending onto the river.

4.4.3   The plume seen from Westminster Cathedral descending to the
ground. The bluish colour is characteristic of any cloud of very small
particles less than about 0.4 μm which is the short wavelength end of the
visible spectrum. This is known as Rayleigh scattering and is the
mechanism which makes the clear sky appear blue (see Scorer 1997) the
scattered intensity being inversely proportional to the fourth power of the
wavelength. In this case the particles are of SO_3 and they are not captured
by the falling water drops with the SO_2 because of their small size.

Professor Brunt said that this downward motion happened because the chimneys could be seen obviously not to obey his two and a half times rule. The eddy motion which the plume made visible showed that a more likely explanation was that it was sinking because of negative buoyancy. The occupants of the blocks of flats on the opposite side of the river complained that they were frequently engulfed in whiffs of the plume which had been lowered in this way; they were pacified by being told that they were getting very cheap hot water which was being piped under the river as waste heat from the generators.

This making use of waste heat fascinated the Westminster City councillors (so a Battersea engineer told me); but actually there was a small generator in one corner, not one of the main four, which was supplying the 'waste' heat. The heat required by the flats was only a tiny fraction of the supply of waste heat from one of the main generators, and rather than complicate the main cooling system and lead off a small amount, it was a more sensible arrangement to provide it all with this tiny set in the corner. That fact was enough to shock the environmentalists who thought that the Battersea station was designed and operated on the best and latest (for them) scientific principles for economic use of fossil fuel, and so the reality was tactfully kept confidential.

To return to the content of the chimney plumes: the engineers concerned with the design said that the cloud of the smallest water droplets in the plume as it emerged from the chimney was not 'splashover' from the 'rain' of much larger drops used to dissolve the SO_2, and therefore it must have been caused by the loss of heat through the chimney stack wall. On the contrary I suggested that by the time the flue gases had reached somewhere in the middle of the rain chamber they had already been cooled down to their wet-bulb temperature, and from there on upwards they were losing *sensible heat* to the raining water which had been introduced into the rain chamber at a lower temperature than the wet-bulb temperature of the flue gases, and this final stage of passing among the raindrops caused the condensation of cloud on the millions of CCNs which are produced in any fuel combustion process.

Conditions were therefore set for bringing the plume down to the ground close to the chimney as the cloud of tiny droplets was evaporated by mixing into the ambient air.

There remains the question why was the washed plume visible and coloured a pale blue? See picture 4.4.3. When the multitude of condensed water droplets had been evaporated by mixing with ambient air there remained the SO_3 droplets, and the small size made the cloud of it acquire a bluish tint.

With the washed and unwashed plumes well separated at Battersea it was easy to study their behaviour visually. From time to time the gas washing mechanisms needed maintenance but the generation of electricity needed to be continued. It was interesting to watch the behaviour of the washed plumes in the same airstream as an unwashed plume. Washing was occasionally stopped for technical reasons, but was sometimes not resumed as soon as possible because the complaints about the plume descending to the ground ceased until washing began again.

4.4.4   Battersea seen from the NE with the washed plume passing between the two north stacks below their tops.

4.4.5   The view from Westminster cathedral during a spell of smog.

The unwashed plumes had significant buoyancy so that their effective stack height was much in excess of the requirement of the two and a half times rule.

4.5   DIFFERENT FEATURES OF THE BANKSIDE PLUME
The newer power station at Bankside a little further down the river from the Battersea station was built to burn oil brought up the river by sea and which contained a greater proportion of sulphur than the coal currently in use. It was therefore logical for gas washing to be installed there also and at the same time to install a four flue chimney, which had recently been shown to get the best advantage from the buoyancy of hot flue gases. Unfortunately this did not cure the cooling by washing problem and the plumes emerged at about 23°C instead of the more usual 90°C, typical of most other stations, and they contained the same cloud of water particles as at Battersea, which made them very white.

4.5.1   Bankside plume descending, as seen from St. Paul's, while on full power.

4.5.2   The plume descending on to Southwark Bridge.

The result was that although when the wind was calm the mixing of the plumes with the ambient air was delayed and they collectively ascended vertically to around twice the chimney height, see picture 2.8.2. The water cloud was always evaporated very quickly unless the air was calm, or already very damp, and so the plume usually descended to street level. It was quickly identified by members of the St. Bartholomew's Hospital Air Pollution Research Unit as 'thin cloud' of SO_3 consisting of very small particles, not droplets. These particles are subject to viscosity and because they very nearly follow the air streamlines around the falling raindrops they are not collected by them. The dioxide gas molecules by contrast are mostly dissolved in the artificial rain.

Because of their small size the cloud of trioxide particles acquires a bluish colour by Rayleigh scattering, like the blue sky. There is always a small amount of trioxide produced when there is some sulphur in the fuel. The main sources of NO_x when fuel is burnt in air are internal combustion engines where high pressures and high temperatures occur together, as in car engines and hot gas turbines of some aeroplane engines. There is much more scope to avoid NO_x production in the turbines because they have a continuous flow of the gases and no spark ignition. Where appropriately designed gas turbine engines are used to generate electricity there is more scope to reduce the production of NO_x because they are not restricted by weight problems, as in aircraft.

4.5.3   Natural gas plume at 14th St. Manhattan. The water vapour in
the combustion products condenses only temporarily but gives a useful
lift and increases the effective chimney height. Unlike the cloud water in
the plumes at Battersea and Bankside the water droplets help to raise the
plume to a higher level in the few seconds before they evaporate again.

4.5.4   Bankside on low load with only one flue operating.

4.6   AVOIDANCE OF COLD EFFLUENT: FLUE GAS DESULPHURISATION (FGD)

In the problem of how to create a plume that will not cause a deposition of acid-loaded snow a chemical reaction is required that will avoid a water-laden effluent. The outline of current FGD is as follows:-

1. The flue gas comes from the generating plant and passes through electrostatic precipitators to remove as much dust as possible, which is quite considerable when pulverised coal is used as fuel. **2.** It then passes through a heat exchanger to lower its temperature to 90°C or less. **3.** It then enters the *Absorber* where it reacts with limestone slurry. **4.** The absorber is a 50 metre tower lined with rubber to prevent corrosion. The slurry is circulated by being sprayed from near the top of the tower, and the limestone is converted into calcium sulphite and at the same time 95% of the hydrogen chloride originally present in the coal is also removed. The limestone was delivered as crushed rock and was converted to slurry in horizontal ball mills. Compressed air is sprayed into the slurry to convert the sulphite into sulphate. **5.** The flue gas now goes through demisters at the top of the absorber which remove any carry-over of wet slurry and gypsum, and then through the heat exchangers which warm it to about 80°C which is warm enough for adequate dispersion. **6.** Limestone is separated from the gypsum, which is then washed to reduce the chloride level and is then placed in vertical axis centrifuges to reduce it to a commercially manageable damp powder. Heavy metals may have entered the system through the coal or limestone and the waste water has to be diluted to render it harmless.

Some sections of this continuous / intermittent flow system need special linings of Glass Flake Vinylester (GFVE) or High Nickel Alloy according to the temperature and chemical content in addition to the Absorber tower's rubber lining.

The system needs its own control room in addition to the electric power output control of the Station. The flue gas emerges as visibly clear from the stack, but a water cloud appears as it becomes cooled by mixture with ambient air. The heat exchangers do not have continuous flow of opposite directional streams; but the flue gas from the precipitators warms it up and the gas from the Absorber is warmed to give it buoyancy before it emerges from the stack.

This system gives very good results; but was described by Lord Brian Flowers as a £100 problem demanding a £1,000,000 solution for the flyfishing industry. It will be seen as trivial economics compared with the problem of reducing the output of CO_2 to avoid excessive global warming. Flowers described what was, after all, only a problem for sporting fishermen. But it was also a problem that must be solved, for the power engineers had in the back of their minds the ultimate problem of the ultimate exhaustion of the world's fossil fuels.

The temperature must be lowered if the maximum absorption of SO_2 was to be achieved: and the effluent must be heated to the maximum to maximise the thermal rise and minimise the adverse effects of the throwaway gas. Clearly the hot gases from the turbines after their greatest expansion must be required to give up as much heat as possible to the effluent which had been cooled to give up its sulphur dioxide. These two gas flows must pass in opposite directions through a heat exchanger. The figures achieved at Ratcliffe coal-fired power station near Nottingham are that on entering the heat exchanger the temperature is 80°C and on leaving the heat exchanger to pass up the chimney, having given up 96% of its SO_2 , its temperature is again close to 80°C.

The SO_2 is absorbed by being passed over Limestone ($CaCO_3$) which it has converted into Gypsum, a mineral which has commercial value in the building industry as a base for wall plaster.

5

Environmental costs and benefits: a matrix for comparison of sources

5.1 THE METHOD

Although the problems of air pollution have been with us for many centuries it has usually happened that most remedies to reduce the harmful effects have not been sought out and applied until they have been clearly seen to be urgently necessary. Methods of domestic heating and profitable industries using cheap fuel have provided comfortable living and a variety of wealth producing industries have all seemed to represent progress. In the early days of growing industries the worst pollution seemed to be located in a few places and only when the polluted areas became large and began to join up with each other did the pollution seem to represent an enormous cost. Only then did it begin to seem that the cost of preventing the pollution might be well spent.

The way to burn bituminous coal efficiently at home was well and widely understood in the year 1900 but it remained common to burn it in open fires and it was normal conversation to speak of the heart-warming nature of an open flaming fire without reference to the misventilation and inefficiency which it caused. At the time of the great urban smogs the combustion of the fuel was usually very inefficient and it gradually became obvious that efficiency in the use of fuel was often the same as minimising the creation of air pollution. After all, smoke is usually unburnt fuel; and that realisation was the step to making people want to abolish pollution. Efficiency was good economics.

After the smogs began to be prevented the realisation that we had never really needed to suffer them came very slowly. The 1956 Clean Air Act was the realisation of much that had already been used in advertisements for domestic closed stoves burning suitable solid fuel. It was before natural gas became available from the North Sea. Also the emphasis on the abolition of smoke made it clear how easy it was to make the air look clean without reducing the amount of SO_2: but the smoke had adsorbed a large amount of that harmful gas and was a major agent for carrying it to the site where the harm occurred.

The method now to be described was designed to anticipate the legislative requirement and defines categories of fuel burners which the law was likely to require to reduce their contribution to the smog. In the 1950s the categories of fuel users were already defined in publicly available information about the consumption of coal, or coal equivalent, as listed in Table 1:-

1. Domestic	4. Industry (chimneys below 200 ft)
2. Road Traffic	5. Collieries
3. Steam Trains	6. Power Generation & other v. tall chimneys

The definition of the fuel users
---------- Table 1. ----------

This simple classification helps in the choice of effective legislation.

5.1.1   Black smoke – easy to ban, because it represents inefficiency. The potteries in Haworth, West Riding, in the 1950s. A common legacy from the era of cheap coal.

There was another category in the original list of Table 1– *Others* –; but it was clear later from the conclusions drawn from the matrix that its exclusion would not significantly affect the importance of the others relative to each other. Anyway the numbers to be chosen for the matrix depended on the concept of a typical member of the category.

When the paper introducing the Matrix idea was presented at a meeting of the Institute of Fuel (Scorer 1957), the discussion turned mainly on the structural differences of various communities within the nation; e.g. Bury and Bolton (industrial) contrasted with Bournemouth and Brighton (seaside). But those differences would not alter the contributions a house or a train was making to the general pollution. The criticism was avoided by introducing the concept of 'black areas' which had already been defined by the national survey of long term SO_2 pollution.

The implementation of the Clean Air Act of 1956 was left to the Local Authorities, who were required to declare smoke control areas in which, after a specified date, it would be illegal to burn other than approved smokeless fuel or to emit black smoke other than for a certain brief period of time for lighting up purposes only.

In order to make this effective the user of fuel was required to use only approved grates or stoves, a large proportion of the cost of which would be paid by the local council which in turn would be subsidised by central government. Every domestic user was visited and advised by the local public health inspector: and the transition to the absence of domestic smoke was very effective in the next few years as the Local Authorities extended the smoke control areas, and were encouraged by frequent public reviews of national progress reported to the Clean Air Council.

Many people, seeing the benefit of smoke control but not yet placed within a smoke control area, obtained the appropriate heating and cooking devices without waiting for the local authority to include their area. It was later suggested that the subsidy had therefore been unnecessary to achieve the transition to smokelessness; but that argument ignored the fact that the declaration of smoke control areas always began in urban centres where the pollution was worse and the householders were usually less affluent and harboured old ideas about the inevitability of pollution being a sign of progress. They would probably have been the last to make the change if it had been at their own expense, instead of being in the first street to be cleaned up. Anyway this is only an echo of the familiar old argument (never abandoned when argument is invited) about the administrative cost of means-testing so that public money should go only to those who really need it.

5.2   MATRIX DETAILS

To complete the matrix, weighting factors had to be applied to each category. Thus:-

Power Stations were rated by the factor 1, and the others rated by comparison in Table 2, (below) with the tall isolated chimney and efficient combustion in mind when making the comparison:

A. Closeness to a main site of the harm due to their pollution.
B. Chemically damaging nature of the emissions to the atmosphere.
C. Concentration into adverse weather – by season, time of day, etc.
D. Height and temperature at emission, which could be redesigned.
E. Effects on sunshine, humidity or other factors such as shadows.

Comparison factors for various sources in Table 1.
---------- Table 2 ----------

The Table 3, below, lists the numbers applied in respect of the mutually independent effects listed in Table 2 to each of the categories of fuel users listed in Table 1.

Category	A	B	C	D	E
1	2	1.7	3	10	2
2	4	1	.6	16	1.5
3	1.5	1.8	.6	13	1.5
4	1.5	1.4	.6	3	1.5
5	1	1.6	1	4	1.3
6	1	1	1	1	1

The matrix of numbers indicating how much worse the categories of fuel users were as pollution sources than an efficient power station with a relatively tall chimney, each using the same amount of fuel (e. g. per ton of coal equiv.) energy.
---------- Table 3 ----------

The pictures opposite illustrate situations to be had in mind when arriving at the numbers in the matrix.
For the subsequent calculations the following were also required:
F.   Amount of coal equivalent burnt annually in each category.
S.   Total product of the independent factors A B C D E.
T.   The product F×S. which gives the relative part of the total damage by pollution due to each of the categories 1...6. It guides the legislator to designing restrictions, as set out in Table 4 (below), to make an effective Clean Air Act .

5.2.1   The steam train firing yards at Chalk Farm, Euston, London.

5.2.2   Black smoke from a hospital in Liverpool showing the disruption of the plume by urban turbulence but remaining within the cone as described in Chapter 3. The combustion is very inefficient and the chimney is virtually 'domestic'.

5.3 NUMERICAL ANSWERS

The paper presented to the Institute of Fuel (referred to in the previous section) was actually published immediately after the passing of the Clean Air Act, 1956, into Law but the nature of the factors had been made a major subject of public discussion, and the paper was a process of mathematical model making to illustrate, and justify, the rationale of the Law. *In discussions many arguments were pursued but no seriously different numerical components or preferred type of legislation was proposed.*

The relative total damage, T (=F×S), by pollution is displayed in Table 4. The damage per ton of coal equivalent consumed is then obtained by dividing up the total damage, at that time often quoted to be £500 millions, in the proportions of the factor T (in the third numerical column). Thus the addition of all the numbers in this column (i.e. 8847) represents the number of damage units caused by air pollution due to the use of fuel. The numbers in the column F give the millions of tons of fuel consumed and the total represents the total national fuel consumption namely 151 million tons (coal equivalent). Therefore 151m × 7548/8847 represents the contribution of category 1 to the total of £500m.

Fuel user Category	F	S	T	Cost of damage in £ per ton
1. Domestic	37	204	7548	11.5
2. Road Traffic	10	58	576	3.3
3. Trains	14	25	354	1.4
4. Industry	44	6	250	0.32
5. Collieries	10	8	83	0.47
6. Electricity Generation	36	1	36	0.03

This table emphasises the relative importance of the different kinds of pollution source. The law which emphasises the importance of getting rid of dark (combustible) smoke is applied to all categories although the application to domestic premises was emphasised because it produced the most clearly perceived result.

---------- Table 4 ----------

After the domestic burning of fuel the next most important pollution problems were seen to be road traffic and steam trains. Since the time of the 1956 Clean Air Act steam trains have been completely replaced first by diesel propulsion and are now well towards a final change to electric power. Road traffic, however, has become equipped with noticeably cleaner combustion. But, and this is not a negligible 'but', at a conference in

February 2000 the representatives of the car making industries warned that in the nature of the internal combustion engine the ideal of quite clean exhaust can only be approached asymptotically towards zero pollution. That fact alone, when properly understood, will mean that improvements will not be expected after about 2003. And by then the sulphur in car fuel will probably have been almost completely eliminated.

The number of vehicles on the roads has about doubled during the last two decades and it appears likely that the main traffic problem will soon be vehicle congestion, which is a main cause of serious vehicle pollution. The doubling (and more) of the proportion of CO_2 in the atmosphere seems to have resulted because of the acceptance by politicians of false arguments that the alleged warming is not taking place – such as:

1. CO_2 is not the most important greenhouse gas, which is water vapour, the atmospheric content of which we cannot control, and which will have a positive feedback if the warming includes the warming of the sea, and may therefore be the cause of any warming that has already taken place.

2. The meteorologists cannot be certain that the warming that seems to have taken place has been caused by some human activity because they do not know the cause of the Holocene, which is the 10,000 years since the end proper of the last ice age. That is to say that we have had a much longer warm period since the end of the ice age than in any of the previous warm interludes the last of which was the Eemian around 100,000 years before the present (BP).

3. For economic reasons we should not inhibit economic growth in order to conserve fuel or to avoid increasing the CO_2 content of the atmosphere when we know that the climate changes that have occurred might be due to what are called 'orbital' changes in the incidence of sunshine, which means that we have just passed a maximum in the mean heating by sunshine and there will be cooling to a minimum in a bit over 10,000 years time.

In recent years we have witnessed the overriding of international agreements by internal US politics, which reveals a preference of local immediacy over long term reaction to global dangers.

Thus, in order to make use of our knowledge when dealing with problems of great complexity we may use the matrix method both to focus our ideas and discover any weak lines of argument, and to give numerical answers to questions which will be asked as soon as legislation seems to be concentrating in particular directions. After the second world war it was widely thought that sociology/politics/economics could benefit from the more scientific methods used with success in engineering. How could the senior administrators know what plans they should approve? The evolution of society could benefit from planning, rather than letting things happen in an anarchy which might or might not produce the outcomes confidently imagined. Perhaps we might benefit from a more democratic making of decisions; and might we not make better use of public 'opinion' surveys? It could be said that we can now apply the message of the work of Isaac Newton because we have computers to work out the implications of the mathematical theories about things. There is no widely acceptable scriptural philosophy or religion able to provide effective discriminatory guidance, for the scriptural language does not deal with modern problems generated by technology. So it is not surprising that the marketplace has been allowed to dictate many of the decisions, and planning has been given direction by cost/benefit analysis, because it expresses the situation in numbers.

As far as cost/benefit analysis is concerned we note that in order to make the problem a numerical one monetary prices or values are rather arbitrarily assigned to all the components involved, usually as a result of a verbal discussion or argument not very different from the process of giving numerical values to the quantities A,B,C,D,E in our matrix.

The original mathematical modelling designed to make use of the well tried and agreed laws of mechanics and physics has to make use of physical coefficients which may not be well determined for conditions outside the range possible in the laboratory or for cases of turbulent motion or the rate of reaction in the case of photochemical changes. When the conditions become too complicated for the agreed equations to be used in obtaining a numerical answer they are represented by a much simpler model, and, of course, the making of that model, though guided by mathematical convenience, is usually a judgement of the convenience it provides as a representation of the real mechanisms.

Thus, the matrix method, like the mathematical model, is a crude representation of reality by numbers which the exact laws require to be used. The arguments used in both cases are a process whereby the participants become appreciative of the roughness of the calculation.

A significant result of extensive computerisation in recent years has been that the name *planning* has been given to an administrative process of facilitating whatever has been predicted by computation to be more or less inevitable. It has often resulted in wishes being given no numerical role in planning; but their vote has been assumed to be determined by cost/benefit analysis. By the actual shortness of the market's foresight the future extent of the decisions and plans has been determined. Too much foresight offers rivals too much time in which to be actively awkward. In the context of the clash between growing population and the shrinkage of unexplored, but hoped-for, resources it has been too easy to avoid thinking about what we are determining inevitably for later on.

At the time the paper on this matrix method was presented (Scorer 1957) the total cost of air pollution damage was recognised as no more than a crude guess. The overall total was chosen because it had been bandied around and to use it in this case put it in 'the literature'. The study indicated that a ton of coal burnt in a house did about 500 times the harm done by a ton burnt in a power station: which was not as shocking a judgement as it sounds if we think of a power station supplying the energy requirements of 2000 houses. Thus it was also said by way of comment that there was not enough emphasis on the beneficial effects of high power station chimneys. This, and the fact that those for the power stations being built would be very much higher together with the fact that they were being planned to be sited far from big population centres and to have multi-flue stacks were all designed to reduce the cost due to the pollution. The only effect so far proposed for traffic fumes is to apply the force of a higher cost of fuel and for the car makers to achieve a higher standard of fuel efficiency, probably based on cleaner fuel.

5.4   ACID GASES: ACID RAIN

But the most talked about criticism of the Clean Air Act was that it seemed to do nothing about SO_2, suggesting that the exercise would have been futile. However it was argued that the effect of that gas on public health was made very much worse by the presence of the smoke. The public was made aware of synergistic effects and that the removal of smoke was easy to achieve. (The same claim was not made for the case of tobacco smoke! The effects of small particles are considered in section 7 of this chapter.)

The Act provided a procedure for determining the height of new chimneys according to the proposed rate of emission of SO_2, and this was to be applied by the local authorities. It was an attempt to define by the use of diagrams the height which would ensure that the maximum ground level concentration of SO_2 would be within acceptable levels. It is noteworthy that it avoided stating the formula used in arriving at the diagrams because there was no agreement on what such a formula should be. Thus for that relatively simple problem there was no generally accepted formula. The subject is considered in Chapter 2. Diagrams were used rather than mathematical formulas because a formula laid the matter open to legalistic argument if anyone chose to flout the regulation. A formula has to be justified, and the range of applicability can be argued about almost interminably. In the event no objections to the procedure with the diagrams have been received.

To the problem of how to reduce the emissions of SO_2 has been added, recently, the much more severe problem of reducing the emission of CO_2. Both these problems have been reduced in France as well as in The Czech Republic and some other East European areas where Lignite (brown

Coal) is plentiful and has more than twice the sulphur content of British coal, by using nuclear power to generate electricity. Another important message from this study is that there is no theory that dispersion can necessarily be satisfactory. Pollution may often be concentrated where it is deposited (Scorer 1994). This was demonstrated by deposition of radioactive dust from the explosion at the Chernobyl Nuclear power station. It was clearly measured in many distant places such as the western parts of the British Isles, Finland, and Italy. The showers by which it was deposited were simply a natural and common mechanism for cleaning the atmosphere, and they drew into their base the contents of the mixing layer around them. We are aware of the power of this process in the case of the desert locust swarms: for the swarm is concentrated in an area where convergence produces the rain which is deposited in showers, and this causes the growth of vegetation and provides food for the hoppers which hatch out as a result of the egg laying when the swarm is concentrated.

5.4.1   The relative deposition of pollution from the Chernobyl power station several days later in Great Britain was highest where rain showers occurred which was mainly in the mountain areas (See Scorer (1997) Sec. 10.5).

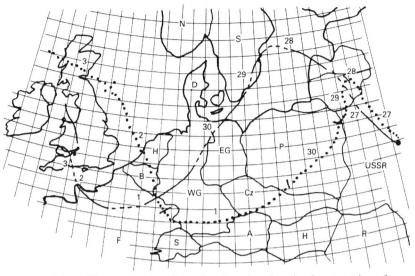

5.4.2   The total deposition of radioactive dust in the countries of Western Europe.

The accumulation in the snow of the mountains of Scandinavia of a winter season's acid rain from industrial parts of Europe is not released gradually as the snow is melted in the spring. When cloud water freezes into snow crystals which aggregate into snow flakes the dissolved acids or salts are left on the outside of the crystals. When the first crystals melt the water dissolves the acids or salts and washes off a good three quarters of them in the first melt.

The consumption of fossil fuels began in earnest around 1750. It is also probable that in the early years of this Industrial Revolution the smoke it produced caused a loss of sunlight by scattering the sun's rays back into space. This would cause a cooling of the lower atmosphere, which would merely have postponed the end of the so-called little ice age which had begun in the 14th century. Cooling at the bottom of the atmosphere has very little direct effect on the air above so that adding a thicker layer of smog would not much increase the cooling of the atmosphere as a whole when fossil fuel began to be exploited in earnest a quarter of a millennium ago.

Also the disintegration of the large molecules of the chlorofluorocarbons by UV sunlight in the stratosphere has recently caused a significant destruction of ozone in the stratosphere, which lowers the equilibrium concentration of ozone. Ozone is a major absorber of UVB at wavelengths less than 0.3 μm Therefore it is feared that more intense UVB at the Earth's surface may cause more widespread skin cancers and possibly also cause harm in other life forms which have evolved during a period of much lower levels of that radiation at the surface.

But in the 20th century the increasing emission of carbon dioxide, which was only being absorbed in the ocean at what may be called a geological rate, began to slow the radiative emission of energy into space and the greenhouse effect began to take hold. It had been thought in the mid 19th century, that there would be a greenhouse effect and when it did not seem to happen it was assumed that the carbon dioxide was being absorbed into the sea; for there was already about 80 times as much in the sea as in the atmosphere and it was argued that whatever mechanisms put that much there would probably cause the absorption of a bit more for quite a few years.

It was only when the long term rise in CO_2 at an observatory in the Hawaiian Islands, far from forests, plains, and dense populations (and other causes of departures from the average) became well known that urban, seasonal, and other variations were understood, that any confident prediction was possible.

Year	CO_2 ppm	Temp °C
1960	317	14.48
1965	321	14.98
1970	325	15.02
1975	331	14.92
1980	338	15.18
1985	346	15.09
1990	354	15.38
1995	361	15.39

Global changes in the average concentration of CO_2 and average temperature derived from global models used in weather forecasting.
---------- Table 5 ----------

The most recent figure is in excess of 400 ppm.

5.5 NEW PROBLEMS: STRAW BURNING, BACKGROUND POLLUTION

A recent problem which came and, as a result of public protest has been dealt with, was the burning of straw after the grain harvest in Britain. This produced extensive areas of smoke throughout the countryside, often severely reducing the visibility on the roads. It was discovered that the exact position of these fires could be identified in pictures made by meteorological satellites, and so a record has been made and is in the archive. If a complaint were lodged the offender could not prove that it could have been someone else's fire that made the smoke. Such was the protest that in a year or two the farmers were saving the straw for some useful purpose. Some were changing to growing rape seed instead of cereals.

5.5.1   A satellite picture after the grain harvest in Southern England with a bright white spot at any position of local high temperature (700 K). Detail can be found in Scorer (1997) (see Hot spots).

5.5.2   Smoke rising from a recently lit straw fire near Lincoln.

5.5.3   A view of the city of Exeter on a summer's day - no problem.

5.5.4   The same view in winter with a light wind. This pair of pictures
reminds us of the common feature that the density of the
background air pollution may vary by a very large factor with
the time of day, from day to day, and with the season through
the year or decade, regardless of considerations of a particular
source of pollution..

5.6   BLACKENING OF SURFACES: TRAFFIC EXHAUST

Surfaces may become blackened in the course of time while adjacent surfaces remain clean. It appears that this may be due to a combination of sunshine, wind, and rain keeping some areas preferentially clean, for they can often be cleaned by a spray of clean water. But this is not the case in Oxford where the stone chosen 2–300 years ago for College buildings becomes softened by (presumably) acid pollution because the detail of the stone carving has been progressively washed or scrubbed away so that it becomes unrecognisable.

5.6.1   Carvings of faces destroyed by pollution over time.

5.6.2   Part of the north side of Exeter Cathedral where the parts which receive the least sunshine have been completely blackened. Is it because they dry out after rain more slowly? In some places it is blackest where water would run down the vertical surface, but there is no obvious consistency in that. Some blackened surfaces seem to thrive on remaining damp after rain; as if there were a biological component in the black material.

5.6.3   Typical thermal precipitation of black particles from the warmest air above a hot pipe on to a cooler surface in a basement room in central London.

5.6.4   A shadow on a flying buttress of the chapter house of Lincoln Cathedral which comes into the shadow of the main building at the time (mid morning) when it would otherwise receive the sun's direct rays.

5.6.5   Sea salt spray made airborne from the whitecaps, at well-named Saltburn, in north Yorkshire.

5.6.6   Trees close to the coast in the Isle of Wight leaning away from
the sea because of the deposition of salt on the sea-facing side which
prevents growth there.

It was possible to look into the future with the aid of the numbers in
Table 4. of section 5.3 above. Specifically we saw the prospect that the road
traffic population would certainly increase and would need to be watched for
rising levels of whatever was in the internal combustion engines' exhaust.
Already in the 1950s, traffic was the second largest polluting category and
some people feared both benzene and lead in petrol as dangers to health.
Because their pollution is mostly emitted in urban areas close to the ground
it was not going to be easy to dilute and disperse. To those who believed that
others would inevitably follow where the USA was leading, the kind of
scene depicted in California under a haze where cars were the largest
polluter, following the USA in Europe and Japan seemed inevitable.

5.6.7   Typical haze made by cars in California becoming more
         common not only in the richest countries but also in places
         typified by Mexico city, Athens, New Delhi, Santiago (Chile),
         Nairobi, and areas of industrial and population resurgence in
         China in spite of the wide use of bicycles.

5.7   THE PM$_{10}$ PROBLEM

There have been several very simple studies made, particularly in the USA, searching for correlations between human mortality and the levels of PM$_{10}$, which is defined as the mass/m^3 of particles having an aerodynamic diameter less than 10 μm. It is clear that the measurements of particles within this range and their size distributions do not discriminate adequately between sizes of essentially different behaviour. It is hoped that the following sizes will be used when PM$_{10}$ is measured, because the names indicate how the human body deals with them when they are inhaled:

Inhalable Nasopharyngeal indicates diameter less than 2 μm but not having large enough fall speed to avoid being inhaled. Particles larger than about 10 μm would mostly be removed from the air breathed up the nose, by sedimentation like household dust. Those which are inhaled would usually be mixed into the mucus and swallowed harmlessly into the stomach e.g. windblown dust, sea spray. Such particles would usually be removed from industrial emissions by a water spray or cyclone centrifuge.

Inhalable Thoracic Typically in the size range 0.1–1.0 μm. If inhaled these particles would be mixed into mucus and transported by the cilia up to the throat and swallowed. Because of their small fall speed they could have a lifetime of days in the air but could be deposited as fine dust in the calm air of a closed room.

Respirable Alveolar Smaller than 0.1 μm i.e. Small enough to be carried into the alveoli where CO_2 is released from the blood and O_2 is absorbed. Thus these particles may be breathed out again or dissolved into the blood in the lungs, or simply deposited and accumulated in the lung, possibly leading to emphysema or cancer in the lungs. This has been common among coal miners and cigarette smokers. The blue exhaust smoke from cars which are burning oil in the cylinders, or smoke from a cigarette held in the hand in calm air and looks bluish, and is in this size range.

Smoke exhaled by a smoker does not appear blue because the size of all the particles has been increased by condensation in the moist air in the bronchial tubes and lungs. Particles in the two smallest size ranges may coagulate to become larger with time and then be deposited like dust.

In recent years the emission of air pollution has been greatly reduced so that whatever is contained in motor vehicle exhaust is now subject to the most intense scrutiny. There was considerable worry about the increased popularity of diesel cars because heavy diesel engined trucks and trains had habitually produced dense black smoke when fully accelerated, either when starting from rest or going up a steep hill. It is interesting that this was reported by Minnaert (1954) in his book *'Light and colour in the open air'* and is repeated without qualification under his name (et al 1993) in a newer revised translation entitled *Light and color in the outdoors*. For many years this black smoke has been a cause for complaint, especially in connection with some buses and taxis in metropolitan areas.

Since the growth in numbers of diesel powered private cars and (in Britain, the stricter regulations for routine tests at time of licence renewal) the development of much improved small diesel engines, in the near future under European regulations these are expected to have very much cleaner exhausts.

The position now is that car manufacturers are confidently expected to produce vehicles which will cause at worst an exceedingly small risk to people in city streets; and from about the year 2004 European standards will be such that in future it will probably be too expensive to achieve any further reductions in particulate emissions as we approach the asymptotic limit of zero pollution. The sulphur content of diesel fuel will be at a very low level and the main concern will be to keep the NO_x low so that summer production of secondary O_3 will be low.

Thus the main urban problem will be caused by old vehicles. The planned production of small diesel engines is very large to coincide with a serious effort to reduce production of CO_2 in line with international agreements intended to reduce the global warming effect. As long as internal combustion engines are required for transport it will have to be appreciated that the perfect ideal of no air pollution can only be approached asymptotically, but never actually achieved.

5.8   THE PM_{10} PROBLEM TURNS MEDICAL, BUT WHAT IS ITS PRIORITY?

So far no causal mechanism has been proposed or discovered which explains how death becomes more likely following the PM_{10} level being higher than usual (Schwartz 1994 i); but there is an apparent relationship which increases the number of deaths by (very roughly) 1.5% for an increase in PM_{10} by 10 $\mu g/m^3$. This very rough result first came from a total of only 4 cities during part of 1985–89, the PM_{10} in the correlation referring to the previous day or to a 5-day or 3-day mean. Thus to get results which can be repeatable and meaningful much more detailed data sets will be required. A causal relationship might possibly emerge. The results so far (e.g. Schwartz 1994 ii) indicate that the morbidity associated with PM_{10} are due to the alveolar-sized particles (i.e. $PM_{0.1}$) which aggravate the respiratory systems of people with already existing respiratory difficulties. The idea that this kind of air pollution can increase mortality seems to be based on the experiences in the 5-day 1952 London smog and other similar severe urban smogs.

Reports of this kind are not very helpful because the nature of the particulate air pollution is not properly specified, nor are the complaints recorded as more detailed than morbidity. Examination by doctors, hospitals, and so on reveal that the patients who died were already at considerable disadvantage to the extent that the mortality referred to the minority who would be likely to meet their last breath quite soon anyway. The situation is described by Maynard (1998) and it arises because correlation cannot be taken as a direct cause and effect relationship until we know the mechanism of causation. An important obstacle to research into the causation process is that the process cannot be investigated by surgery leading to recovery without which death would have been inevitable. Those who die can be subject to post-mortem examination and there can be no doubt that death occurs only in those who are already suffering the symptoms likely to cause heart failure through ischemia or atherosclerosis or thrombosis very soon anyway. Surgery to study the reaction of the body to the invasion of the alveoli by typical alveolar-sized particles has to be undertaken in animals before determining whether continual frequent exposure to episodes of particle air pollution could lead progressively to death of average citizens or whether continual exposure to occasional episodes can be followed by normal recovery each time. Indeed it is probable that the average person is put in that latter category as we approach ideally clean air asymptotically.

Undoubtedly cigarette smoking causes lung cancer and this has caused difficulty in many epidemiological studies. It has been claimed in many papers that confounding causes, such as smoking, can be taken into account by such devices as parallel studies of a similar kind with non-smokers. What is usually revealed in these cases is the terribly addictive consequences of smoking. Other people can be prepared for the death waiting for them in the next severe particle pollution episode by other causes than smoking. Such factors as pneumonia or even severe asthma, sex, age, educational level, and type of occupation have been suggested as possible confounding factors which are eliminated from the correlations with the daily measurements of PM_{10} because they have no relation to the daily rhythm of the observations. This may well be true of the morbidity, but the mortality may represent the dominance of smoking as a major cause of severe respiratory malfunction and readiness for death in the circumstances under discussion here. Thus it would be argued that smoking might be the cause of mortality following episodes of high particulate alveolar air pollution.

Nevertheless, in the six cities study in the USA the figures were said to be 'adjusted for sex, age, smoking habits, educational level, occupation and body mass index ...' to the point of obsession; then we read that 'As far as can be ascertained on the scale of this study, the excess mortality was confined to cardiopulmonary causes and lung cancer.' In view of the pollution history of the past few decades why do we need to look any further for the 'causal relationship mechanism'?

5.9   THE UTAH VALLEY STEEL MILL

The PM_{10} studies in the USA included a very interesting contribution in UTAH (Pope 1989,1991). This refers to an examination of hospitalisation of children for respiratory illness in three adjoining counties: Utah county, Cache county, and Salt Lake county, which were said to have similar housing and demographic patterns. Notably, this study did not concern itself with adult mortality which, as we have just seen, occurred in the other studies of particulate pollution exclusively among people who had developed a lifelong pattern of bronchial ill health. But in the mid 1980s the rate of hospitalisation for respiratory illness in children in Utah county was about twice as great as that in the two adjoining counties.

Utah county had a steel mill in a valley subject to temperature inversions. In August 1986 the mill was shut down for 13 months because of a strike; during which time the relevant hospitalisation rate in that county became indistinguishable from that in the other two counties. But when the mill reopened the rate rose again to double that in the other two. It is thus reasonably concluded that the steel mill emissions were the cause of the doubling.

In many other cities similar association of acute respiratory symptoms are reported by Schwartz (1994), and Pope (1991) associated school absences and increased need for asthma medication following increased air pollution. There are therefore misgivings about the probable increase in the use of diesel engines in the coming decade with a probably increased number of cars. Obviously the levels of $PM_{0.1}$ should be measured and carefully watched, because of an undertaking by governments to decrease the world-wide production of CO_2 and the probable increase in unusual and more violent weather phenomena of global warming. This latter needs much more attention in terms of human mortality and long term costs than may be due to particulate air pollution. We need to be aware that decisions based on so-called cost/benefit analysis made by present day bankers and politicians is likely to keep eyes focused on the immediate profits of the car industry.

They will also continue to regard significant decrease in the rate of population growth as being beyond human control because of the high levels of illiteracy. Thus it is preferred to leave the population problem to the growth itself to make the necessary impression on people. There is a serious need to get people asking 'What are we doing to maintain, in the long run, the major aspects of our civilisation'?

6

THE TECHNICAL DICTIONARY

Absolute Humidity

This is defined as the mass of water vapour, in grams, mixed into one kilogram of dry air, and is expressed in g/kg.

The humidity cannot be greater than when the vapour is saturated. If the air is cooled adiabatically by being carried upwards to a lower atmospheric pressure, when saturation is reached the temperature is cooled to its wet bulb temperature.

Absolute Zero of temperature

This is the temperature at which the molecules of a substance would have no energy of thermal agitation, and would possess no energy in the form of heat. On the Centigrade scale it is $-273.2°C = 0$ K The absolute scale of temperature is denoted by K (Kelvin).

Adiabatic Equation

In considering the changes in the condition of a parcel of air in the atmosphere we use adiabatic changes as a reference for more complicated ones. Adiabatic means 'without passage of heat across the bounding surface of the parcel'. There is assumed to be no transfer of heat by radiation or conduction into or out of the parcel. However energy may be transferred by mechanical work being done on or by the boundary to compress it or allow it to expand.

If dQ is the energy change in a parcel of unit mass and T is its temperature then

$$dQ = C_v \, dT + Apd V \qquad (1)$$

where C_v is the specific heat of the gas at constant volume, A is the mechanical equivalent of heat and pdV is the work done by the gas in expanding its volume by dV at pressure p. Using the calculus method for small changes we neglect second order quantities, in this case the pressure might change during the expansion dV.

In a change at strictly constant pressure we would write

$$dQ = C_p \, dT \qquad (2)$$

where C_p is the specific heat at constant pressure, and would include any work done and heat lost by the expansion dV.

The gas law for a perfect gas, which is a very good approximation for changes in the atmosphere, where the volume occupied by the gas molecules is a very small fraction of the total volume and the molecules are regarded as the 'material' of the gas, is

$$pV = RT, \quad \text{or} \quad p = R\rho T \qquad (3)$$

where R is the gas constant and ρ is the density, so that the change

$$pdV = RdT - Vdp = RdT - \frac{1}{\rho}dp \qquad (4)$$

and for the case of constant pressure we can equate the values of dQ in (1) and (2) with $dp = 0$ to give

$$C_v \, dT + ARdT = C_p \, dT$$

or $\quad A = (C_p - C_v)/R \qquad (5)$

which simply expresses the difference between the two specific heats in terms of the mechanical equivalent of heat. We now introduce their ratio, which we denote by the symbol γ

$$\gamma = C_p/C_v = 1.403 \qquad (6)$$

very approximately for the atmosphere which is composed predominantly of diatomic molecules.

When $dQ = 0$ the change is adiabatic, and the change in temperature is due to the change in density during which work is done on the boundary of the parcel to change its volume, and so we write

$$0 = C_v \, dT + (C_p - C_v) \, p \frac{dV}{R}$$

$$= dT + (\gamma - 1)(dT - Vdp) \quad \text{by (4);}$$

and then, using (3) to eliminate V, we get

$$\frac{dT}{T} = \frac{\gamma - 1}{\gamma} \frac{dp}{p} \qquad (7)$$

which means that

$$\frac{T}{T_0} = \left(\frac{p}{p_0}\right)^{(\gamma-1)/\gamma} = \left(\frac{\rho}{\rho_0}\right)^{\gamma-1} \qquad (8)$$

where suffix 0 indicates the starting value before the change. These results may be written as

$$\frac{\rho}{\rho_0} = \left(\frac{T}{T_0}\right)^{1/(\gamma-1)} \quad ; \quad \frac{p}{p_0} = \left(\frac{\rho}{\rho_0}\right)^{\gamma}$$

$$\text{or} \qquad \frac{Dp}{Dt} = c^2 \frac{D\rho}{Dt} \qquad (9)$$

where $c^2 = \gamma RT = (\text{velocity of sound})^2$

Adiabatic Lapse Rate - a well mixed atmosphere

In a 'well mixed' atmosphere any two parts can be interchanged without altering the temperature anywhere, because all parts have the same **potential temperature**. If we apply the hydrostatic equation

$$\frac{\partial p}{\partial z} = -g\rho \qquad (10)$$

and the equation expressing the gas laws

$$\rho = \frac{p}{RT} \qquad (11)$$

to the adiabatic equation in one of the forms (9) we may derive the lapse rate of temperature for the well mixed state to obtain

$$\frac{\partial T}{\partial z} = -\frac{(\gamma-1)g}{\gamma R} = -\Gamma \qquad (12)$$

The value of R is 2.8703×10^6; and for dry air in cgs units

$$C_p = 0.2396, \quad C_v = 0.1707,$$

so that $\gamma = 1.403$ \hfill (6)

and so with $g = 980.6$ cgs, the **dry adiabatic lapse rate** is

$-\Gamma = 9.86°C$ per km.

For practical purposes, and within the accuracy of usual measurements of temperature we may say that in a well-mixed neutrally stratified atmosphere the lapse rate is

DALR $= 1°C$ per 100 metres.

This means that *heights (layer thickness) between isobars on the tephigram* may be estimated as 100 m for each 1°C along the adiabatic line from one to the other.

Aerodynamic Diameter

The aerodynamic diameter of a particle in the atmosphere is the diameter of a spherical particle of unit density that has the same fall speed in air as the particle. This is a convenient way of discussing the particles of different origin, the fall speed being the most important distinguishing feature. There are some particles for which this is not a good distinguishing feature, the most important being asbestos which consists of many very thin fibres and may have very serious effects if it becomes embedded in flesh. It is particularly dangerous if it penetrates into the alveoli of the lungs, or any part of the bronchial system.

Aerological Diagram and The Tephigram

A diagram convenient for displaying the information obtained from a sounding of the atmosphere, from which useful calculations can be made, has T and $\ln p$ (temperature and logarithm of pressure) as rectangular co-ordinates. These are widely used routinely in some national meteorological services. and are called the '$T - \log p$' diagram. But it is considered more convenient to give the adiabatic changes prominent status as a reference rate of change with the same slope everywhere on the diagram, and this is done by using **entropy** as the second co-ordinate denoted by ϕ in the place of the logarithm of pressure. This is called the **Tephigram**. It has the important feature that the energy involved in the transition from one sounding to another is represented in the same proportion by the area between the two soundings in all parts of the diagram. Also an adiabatic displacement from one level to another is represented by straight lines with the same slope everywhere, and these are called *dry adiabatics* because it is assumed that any condensed water is removed and vapour is supplied at the current temperature so as to maintain saturation throughout the displacement.

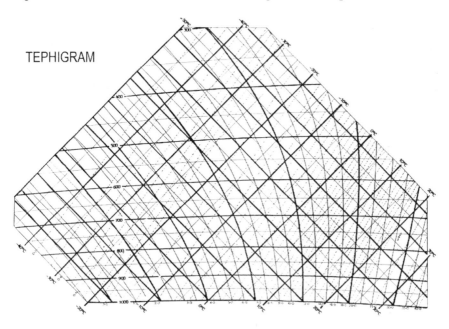

TEPHIGRAM

DIAGRAM 6.1   The $T\phi$-gram

Legend
ALMOST HORIZONTAL STRAIGHT LINES Isobars
STRAIGHT LINES INCLINED UP TO LEFT Dry Adiabatics ⎞ rect.
STRAIGHT LINES INCLINED UP TO RIGHT Isotherms   ⎦ co-ord.
LINES CURVED UPWARDS TO LEFT Wet Adiabatics
DASHED INCLINED STRAIGHT LINES the numerical value of the
Humidity Mixing Ratio is indicated for water vapour saturation.

Also displayed on the $T\phi$-gram (which was devised by Neuhoff and modified for meteorological use by Napier Shaw) are the *wet adiabatics*, which are the curved lines of slope gradually approaching that of the dry adiabatics higher up the diagram. They include in the heat exchanges the release of latent heat when the saturation water vapour mixing ratio changes. This latter quantity is represented by the lines of constant saturation mixing ratio lines which are dashed lines inclined to the isotherms in the diagram.

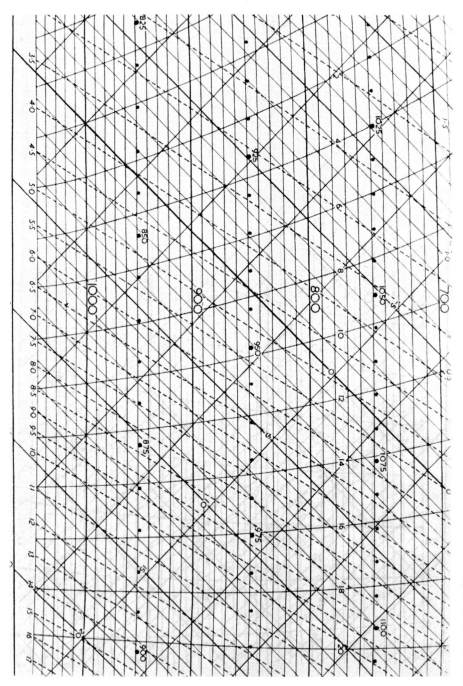

DIAGRAM 6.2   Full page diagram 6.1. (version arranged for ease of photocopying)

Each of these lines is labelled with the appropriate saturation mixing ratio of g of water vapour per kg of dry air, indicated below the bottom isobar of the diagram.

In Shaw's version of the diagram it is assumed that any liquid water released by the condensation of vapour in any displacement in the diagram is removed.

Obviously the starting point, in any series of changes, determines how much water is available for condensation, and in this diagram it is assumed that any water condensed is assumed not to affect the later heat exchanges. Thus the wet adiabatics are not strictly adiabatic because heat energy contained in the water that must be removed or supplied in order to keep the air just saturated, and the lines are therefore often called pseudo-adiabatic. But the errors resulting from this as a representation of how the temperature would actually change are quite small except in the case of large displacements, and are not usually likely to be any larger than common errors of measurement.

Anvil cloud

PICTURE 6.3   An anvil cloud in the Caribbean, seen from a flying boat (*Claude Ronne*). (See also picture 1.13.1)

When convection forms cloud and reaches up to an inversion which the rising air is not buoyant enough to penetrate, the cloud spreads out horizontally to form an anvil. This is characteristic of cumulonimbus but also occurs on smaller unfrozen cumulus to form anvil-stratocumulus.

Bergeron-Findeisen Mechanism

When it was noticed that in clouds composed of supercooled water droplets some of the droplets became frozen, (see **fallstreak holes**) and the freezing rapidly spread outwards from the initial glaciation, it was realised that in clouds of greater volume this could be a mechanism for the whole cloud to become glaciated. At this suggestion it was thought, especially in the cooler higher latitudes that this was THE mechanism by which rain became released from clouds. But to meteorologists with experience in warmer climates it was plain that rain could be generated without freezing being involved. (See **warm rain** and picture 6.26).

Black Areas

There was plenty of observational material about smoke and SO_2 in Britain, and when it was mapped by the *National Survey* the Black Areas were indicated as the most severely polluted, probably containing most of the severely blackened buildings. The first national Clean Air Act, 1956, required the local authorities in the Black Areas to declare smoke control areas without delay, and to follow this with the surrounding areas as soon as possible. The makers of smoky effluent were given subsidies to make the necessary conversion to the use of coal or other fuel without making black smoke except during very short periods such as when lighting up.
See for example picture 5.1.1 (Chapter 5)

Buoyancy

According to Archimedes's principle a body immersed in a fluid will experience a buoyancy force equal to the difference between its own weight and that of the fluid it is displacing. Thus a body of fluid immersed in a fluid of different density is subjected to a body force equal to the difference of the densities of the two fluids.

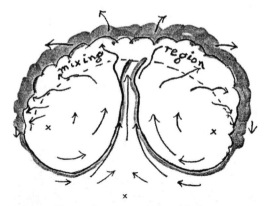

DIAGRAM 6.4   The motion pattern induced by buoyancy.

Buoyant Convection

When two adjacent parcels of air of different density are at the same level (altitude) the denser experiences a downward, and the less dense an upward buoyancy force. This is equivalent to an angular acceleration about a horizontal axis, which therefore generates (horizontal) vorticity.

Thus random **vorticity** is generated wherever there are (random) horizontal density gradients. The vorticity creates **mixing** with the air into which the buoyant air is rising, and this is a form of **turbulence**. The volume of buoyant air increases as it rises and is thereby diluted. See also **thermal** (below).

PICTURE 6.5   Experiment in water tank showing increase in volume: series of four pictures.

Castellatus

This description is applied to towers of cumulus cloud growing upwards above the condensation level when there appears to be no connection with buoyant convection below the cloud base. They usually appear in groups or rows giving a turreted appearance to the outline.

PICTURE 6.6   Castellatus cloud

Cirrus

This name was given originally by Luke Howard (1810) to clouds which look like a curl of hair. They are ice clouds and when they exist in air that is just saturated for ice they may remain with only little change of appearance, often for several hours. When water cloud begins to become frozen the crystals will grow because in water cloud the air is supersaturated for ice, and so the crystals may be seen to fall as they grow. They may later be seen to evaporate if they fall into air which is unsaturated for ice.

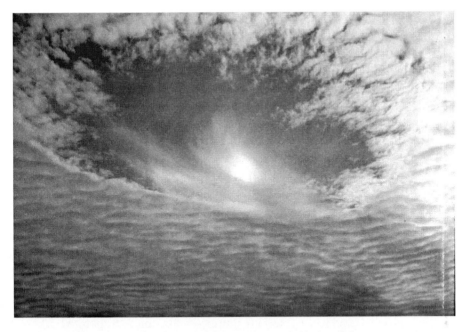

PICTURES 6.7 and 6.8   Fallout of glaciated cloud particles

Cloud Condensation Nuclei (CCN)

If the air is cooled, by whatever mechanism, condensation of the surplus water vapour beyond the temperature at which it becomes saturated requires the presence of nuclei (CCN) or a solid surface on which the liquid may condense, otherwise the air would become super-saturated. In practice some CCN are usually present so that a measurable degree of supersaturation is not actually observed. If there is a dearth of nuclei the number of droplets may be significantly reduced, and the size of the individual drops is increased to take up all the excess water vapour (see **ship trails**).

The **saturation vapour pressure** over an ice surface is less than over a water surface. This means that at temperatures below 0°C, the **dew point** is colder than the **frost point** (q.v.).

Condensation Level

This is the level at which the air becomes saturated as a result of dry ascent. During the ascent the air cools at the **Dry Adiabatic Lapse Rate** (**DALR**) which is roughly 10°C per km of ascent. The wet bulb temperature decreases at the wet adiabatic lapse rate which varies according to the water vapour content of the air.

In air that is descending and contains cloud its temperature increases at the wet-adiabatic lapse rate and the cloud becomes evaporated when it reaches the condensation level (which is now the *cloud evaporation level*) beyond which it warms at the DALR.

Contrails

This is the word for 'condensation trail' which is formed in the water vapour in the engine exhaust of an aircraft in the upper troposphere, or sometimes in the lower stratosphere. They may persist for hours if they are frozen, but otherwise evaporate quickly. Their persistence implies a very slow mixing with the surrounding air, particularly in the stratosphere. See picture 1.7.1. (For full details see Scorer (1997).)

Convection

This term literally means 'carried by the motion (of the air)', which usually means 'by the horizontal component of the wind'. but in contexts which usually make it clear it is used as an abbreviated form of *buoyant convection*, meaning *thermal* convection, which results from heating of the air from below. It is therefore very dependent on the *Diurnal Variations* of temperature and the latent heat due to condensation of cloud.

The **Aerological Diagram** shows the mechanisms quantitatively.

Cumulus Cloud

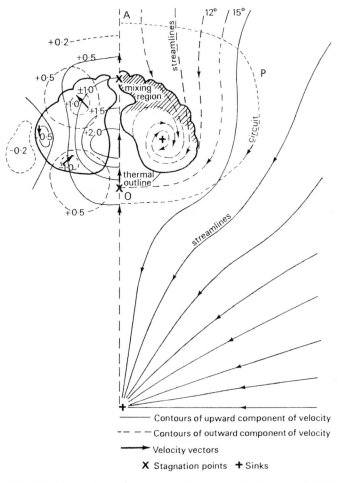

DIAGRAM 6.9   Circulation measured in a thermal (*Woodward 1959*).
In the thermal the numbers represent the ratio of the air's vertical velocity
as a multiple of the rate of rise of the thermal as a whole. Since the *shape*
of the thermal remains unaltered as it rises and mixes with surrounding air
we can represent it in reverse as being stationary in the environmental air
moving into a sink below the thermal marked by a + sign. It shows clear
air rising up the middle into the mixing region at the top.

Cumulus is the cloud form created by thermal convection when the
buoyant air rises above the condensation level.

In a buoyant mass of air the interior air rises to the top and then
spreads out sideways, being replaced at the centre by fresh air from below.
As it mixes with the exterior air, which is unsaturated, some of the cloud

droplets become evaporated and the cooling caused thereby is often enough to generate a downdraught around the rising cumulus tower, the mixed air being colder than the cloud air. Cumulus clouds are continuously disappearing by evaporation around the outside and are usually being replaced by fresh thermals from below. The clear air between cumulus clouds becomes stably stratified at approximately the wet adiabatic lapse rate; and is therefore likely to become warmed by the **subsidence** which occurs in response to the rising air supporting the cumulus. Thus the air above the cloud base may be warmed by convection even though very little cumulus carries any pollution above the condensation level.

Dew Point

If air is cooled at a constant pressure, as in a refrigerator or when the ground cools after sunset, the dew point is the temperature at which the air becomes saturated and dew begins to be deposited on almost any cold enough surface.

For cloud to be formed in air (or in any other gas) when the temperature is decreased to below the dew point, as when it is cooled by adiabatic ascent, condensation nuclei (CCN) must be present in the air but see **frost point**. If no **CCN** are present the air becomes **super-saturated** for water vapour. (See also **Normand's theorem**.)

PICTURE 6.10   Small drops of 'dew', and a large drop of 'guttation' on the blade tip driven out by the higher internal pressure in the warm grass roots.

Diffusion

This is the mechanism by which pollution concentrated in a small mass of air is spread throughout a greater mass. It always produces a dilution of the material being diffused. But if water vapour is being diffused upwards by convection there may be an increase in the concentration of water vapour at higher levels; likewise see **fumigation** for a case of increasing pollution at ground level by downward diffusion.

There is often said to be 'diffusion by the wind' at a source such as a chimney, and many modelling formulas include the factor $1/U$ accordingly. But if the wind is not turbulent, this merely represents the mass per unit length of plume placed into the wind, i.e. *dilution by emission into the wind.*

Dilution

Dilution means the reduction of the concentration of a pollutant by mixing into a greater volume of air. If it takes place as a result of molecular motion it is less than dilution by eddies if the eddies are not simply oscillations as would be observed at a point (referred to as *Eulerian*) in a fixed co-ordinate system through which the fluid is passing. If there is no vorticity the fluctuations may nevertheless be called 'turbulence' (q.v.), but if they are elastic (sound) or gravity waves no mixing occurs. If the motion is observed in a *Lagrangian* manner by following the changes that occur in the surroundings of a specified parcel or particle of the fluid, the fluid may be mixed and enter new surroundings if there is no subsequent reversal of the fluctuations.

Dispersion

To be dispersed means to be carried away to somewhere else. Thus a tall chimney more effectively disperses the effluent gases than a very short one because it removes them from the ground, and to a level at which the wind is usually (but not always) stronger.

Although to carry pollution away is to disperse it, it is important to note that the turbulence generated by the wind passing over rough terrain is very much less effective in dispersing it upwards than the turbulence of buoyant convection. Buoyant air is carried upwards *through* its surroundings by the rings of **vorticity** which its buoyancy generates.

Divergence

When divergence occurs the air spreads out horizontally so as to occupy a greater horizontal area but at a lower level (and atmospheric higher pressure). The warming which results can be discovered by moving the representative points along adiabatic lines on the tephigram.

Drop size distribution in clouds and fog

When condensation of cloud occurs in rising air the size of the droplets is determined by the availability of CCN present. In chimney effluent CCN have been plentifully produced by combustion, and so plumes contribute to air in which cloud is formed, as when the fuel is natural gas, the cloud contains a very large number of droplets and appears very bright (and looks very clean) in sunlight. A relatively large proportion of the sunshine is scattered out of the cloud and often away from the Earth and is lost into space. The strongest incident sunshine experienced on Earth is where scattered sunshine from surrounding growing cumulus clouds is added to direct sunshine.

As soon as air ceases to rise where cloud is being formed the very slight supersaturation is decreased and the smallest droplets, in which the surface tension produces the greatest internal pressure begin to evaporate. Accordingly the largest drops begin to condense more water vapour and there is an increase in the average droplet size and a decrease in their number, while the amount of condensed water remains approximately the same. The total cross-section area of the droplets is decreased and the cloud becomes less bright in sunshine. In general, since pollution usually increases the number of CCN especially of hygroscopic ones which are less likely to become evaporated, pollution tends to produce a cleaner appearance of the cloud.

Details of the classical kinetic theory of gases and its application to the phenomenon of capillarity, and the meniscus or other curved surfaces, is applicable to the evaporation of small droplets. (Gk. meniskos = crescent, dim of Lat. Mene = Moon).

Fallout

Typically this means rain, and equally may refer to hail, snow, graupel (soft hail), or drizzle, but not cloud particles whose fallspeed is at least an order of magnitude smaller. Cloud particles do fall out of cloud bases but usually evaporate before they have significantly altered the appearance of the cloud base. However at temperatures colder than 0°C if cloud droplets become frozen they become surrounded by water vapour that is saturated with respect to ice and may grow rapidly to become fallout particles. If they fall into air whose temperature is above 0°C they have become rain by the **Bergeron-Findeisen** mechanism, which is exploited in cloud seeding attempts to produce wet fallout.

Fallstreak

A visible streak of ice particles falling from a region of ice cloud where the particles have been growing in air supersaturated for ice.

PICTURE 6.11   A fallstreak, particles falling through wind shear.

Fallstreak holes

If an aircraft passes upwards, or downwards, through a layer of cloud of supercooled droplets the impact of a part of the aircraft usually causes immediate freezing. The ice may remain attached to the plane as icing, or may be thrown off and initiate freezing of the cloud which will then be in air supersaturated for ice. The ice particles will therefore grow rapidly and become fallout as a fallstreak, leaving a hole in the cloud.

PICTURE 6.12   A fallstreak hole.

Fog showers and smog hoar frost

If a cloud of supercooled water droplets impinges on a solid surface it may freeze as solid ice onto the surface as **rime**. If the same air, or the cloud droplets, contain fog particles (i.e. is smog) the rime will be dirty. The rime is then said to have been produced by the passage of fog showers. This name was given to dirty rime by Dr. Koldovsky at the Milesovka Observatory where it was common in winter.

Frost Point

If air is cooled below the frost point, as in adiabatic ascent, frost (i.e. ice) is formed on any suitable surface, such as almost any solid or on any ice **nuclei** that may be present such as ice particles or other roughly similar crystalline particles, the best known being silver iodide. In the absence of suitable nuclei the air becomes **super-saturated** for ice.

PICTURE 6.13   Hoar frost in a London smog. 09.00 7 Dec. 1962.

PICTURE 6.14   Dark bands in rime on a vertical rod at Milesovka.

The table below gives the difference between the frost point and the dew point showing it to depend on the temperature.

The dew point has no practical meaning at temperatures below $-40°C$ for pure water because freezing is virtually spontaneous unless it is prevented by an acid or salt being dissolved in the water (see **ozone hole**).

T_d °C	0	−4.4	−9	−14	−19.4	−24.5	−32.2	−40.4
T_f−T_d °C	0	0.5	1	1.5	2	2.5	3	3.5

TABLE 6.15   The difference between Frost point T_f and Dew point T_d.

Fumigation

This term was proposed by Dr. E. W. Hewson to describe the downward diffusion to the ground of pollution from an upper layer in which it has been transported horizontally by the wind from a source a long distance away over the top of slower-moving air. Thus the origin of the pollution is not obvious to those who are receiving it at ground level. This is most likely to happen when early morning convection reaches up to the layer of pollution placed under an inversion by convection during the previous day or night.

PICTURE 6.16   Fumigation: plume formed over night at Mt. Isa and which obscured the air field for 40 minutes that morning.

Greenhouse Gases

A gas which absorbs in the IR wavelengths but is transparent to visible wavelengths and therefore allows visible sunshine (6000 K) to pass downwards and warm the Earth's surface, but absorbs into the air the IR emissions from the Earth (300 K) (see **Planck's emission spectrum**) and so effectively 'acts like a greenhouse' in raising the temperature of the air near the surface.

Such absorbing gases have larger and more complex molecules than the main atmospheric gases – O_2, N_2, A (Argon) – the most important being H_2O and CO_2, which are by far the most abundant of the greenhouse gases in the atmosphere. Also important are CH_4 (Methane) and its derivatives the Chlorofluorocarbons (CFCs), and these are up to 20 times more absorbent molecule for molecule than CO_2. The CFCs are not produced naturally and are therefore a form of artificial pollution. Similar bromine compounds may be produced by the combustion of petrol additives, but may also be created in forest fires.

O_3 (ozone) is very important in the stratosphere because it also absorbs strongly in the UVB at wavelengths less than about 0.3μm.

Hydrostatic Equation

When the air is static there is no horizontal gradient of pressure, and the vertical gradient is described by

$$\frac{\partial p}{\partial z} = -g\rho \qquad (10)$$

This is proved by considering the net force on a unit volume of air to be zero.

Haar (Scotland)

Very low cloud or fog drifting across the coast from the sea.

PICTURE 6.17   Haar near Aberdeen.

Ice Cloud

This is the term used to refer to a cloud consisting of frozen water (ice) particles. It is usual for pure (liquid) water cloud to freeze spontaneously at temperatures below −40°C. At intermediate temperatures up to 0°C water clouds may remain unfrozen for many hours but once freezing begins it usually spreads within the cloud. Such freezing is more likely at lower temperatures, and may take place within minutes at −35°C or so. (see **Frost Point** and **Contrail**)

Inversion

In a sounding the temperature most usually decreases upwards, so that when it increases upwards it is described as an **inversion** (of the usual condition). In an incompressible fluid, such as water is approximately, an inversion is a statically very stable layer, and the same applies in the air if the *potential temperature* increases upwards. As a consequence the word inversion has come to mean a very stable layer in the air; but of course a stable layer means only that the temperature decreases upwards more slowly than the adiabatic lapse rate. In that sense it is used throughout this book.

Lapse Rate

The rate at which a quantity decreases with increasing height is called its Lapse Rate. (e.g. See **Adiabatic Lapse Rate**)

Mixing Layer

This concept is useful in describing the layer into which pollution emitted at the surface becomes mixed under the prevailing circumstances. It is assumed to be well defined, and is not an appropriate concept for days on which it does not have a definite top. This top is sometimes well defined at the condensation level, which is where the cloud base of cumulus is found. Its top is then defined by the **sub-cloud inversion** which may often be seen from the air as a haze top. As long as cumulus continue to be formed this layer is *well-stirred* but this mixing ceases in the evening and the air may become stably-stratified, and become well mixed only some time after sunrise.

It is presumed in most mathematical models of pollution dispersion that a mixing layer exists, and although this may seem to be an extravagant assumption it may be because the pollution is confined that the case is interesting.

In box models which are used to study cases in which chemical changes produce secondary pollution (which was not emitted to the atmosphere in this later form) the mixing layer may be an essential feature.

The mixing layer is often seen to be capped by a haze top which is below the higher mountains in the region, the pollution having been confined by the hills.

Normand's Theorem

During adiabatic ascent the Dry-Bulb temperature follows the dry adiabatic line on the aerogram; the Wet-Bulb temperature follows the saturated adiabatic curve; and the Dew Point follows the saturated water vapour content (dashed) line. All three lines meet at the condensation level.

This facilitates the calculation of the **condensation level** in buoyant convection and of the **Dew Point** in cooling by radiation or conduction if the wet- and dry-bulb temperatures are first known.

Ozone destruction and the Ozone Hole

The UV component of Sunshine causes the decomposition of large molecules in the stratosphere, and also reduces some molecules to atomic state. A sequence of complex photochemical reactions often generates a steady state and includes ozone in the normal composition of the stratosphere. This increases from the tropopause up to about 30 km, where although the proportion of ozone continues to increase upwards the amount per unit volume has decreased from the maximum between 20 and 25 km because the total air density decreases logarithmically upwards. Consequently the region between 100 mb (about 16 km) and 45 mb (about 21 km) is often referred to as the '**ozone layer**'.

It had become normal for a great concentration of ozone to remain in this layer over the winter pole and to endure through the winter darkness when the photochemical reactions which might cause reduction of the ozone concentration would not occur. The same air mass would largely remain over the winter pole in the southern hemisphere winter because of the axial symmetry of the situation in which the cold air was very slowly sinking. Then in October and November as the sunshine spread into this region photochemistry would be resumed.

The increasing use of chlorofluorocarbons (**CFCs**) from around 1960 as their usefulness became better known, was followed by an increase in the total amount of CFCs in the atmosphere because there is apparently no natural process which would decompose them in the troposphere. Consequently they became decomposed in the UVB light of the southern spring in the stratosphere so that there were compounds of chlorine ready to be processed in the ozone layer.

It had already been predicted by Professor S. Rowland in Utah that the NO_x in the exhaust of aircraft flying in the stratosphere would lead to a series of photochemical reactions which would 'destroy' ozone and thereby permit the shorter UV component of sunshine to reach down to the ground through any cloudless sky. This would increase the incidence of skin cancers particularly among light skinned people whose numbers had increased greatly in the 20th century.

The number of aircraft planned to fly in the ozone layer was never large enough to bring about significant ozone destruction because of the poor economic prospects. It was also discovered that there was a significant amount of NO_x present naturally anyway and this was kept at a 'safe' concentration by the fact that the photochemistry was continuously converting it irreversibly into nitric acid which was quickly rained out if it was diffused down into the troposphere.

The CFCs presented a different problem: their decomposition by UV in the stratosphere caused the generation of Chlorine Oxide (ClO) and the detachment of the oxygen atom would react in photochemical reactions with O_3 with the resulting production of O_2, i.e. the destruction of ozone.

At the time there did not appear to be much reduction of ozone as a result of the CFCs in the atmosphere, and it was appreciated that many of the photochemical reactions required a solid or liquid surface on which the reacting substances would be adsorbed, to provide a location for the reactions to take place. That meant that the destruction of ozone would not occur significantly when only gases were present. Furthermore Prof. Rowland pointed out that the dangerous substances ClO and NO_2 could put themselves out of action by forming Chlorine Nitrate ($ClONO_2$); although he was soon the first to point out that unfortunately this new compound would itself be readily decomposed into its components by the short wavelength UV of sunshine!

No great destruction of ozone seemed to be happening according to the theoretical predictions until quite suddenly a very considerable reduction was reported over Antarctica where observations of total ozone overhead had been measured by a team from Oxford using a method invented by Prof. Dobson and currently based in the far south.

This decrease in total ozone measured in a vertical column was called the *Ozone Hole* and its extent was very similar to the 'land' area of Antarctica and was delineated by observations made from polar-orbiting satellites. It has been explained, although it had not been predicted, by noting three requirements: first that there existed an abundance of stratospheric wave clouds over Antarctica produced by the airflow of the winds, which are strongest in the southern winter, in the circumpolar vortex over mountains in Antarctica. Secondly the temperature in these clouds is the lowest to be found in any atmospheric clouds and are close to −90°C, and are produced by a radiation balance between outer space and the surface temperature of the Antarctic ice plateau. And thirdly the appearance of ClO due to the disintegration of CFCs which have been diffused upwards from the troposphere. The only way to prevent this ozone destruction is to reduce as much as possible the global manufacture of CFCs.

PICTURE 6.18   A mother-of-pearl cloud in the stratosphere, which is very cold, but the droplets are not frozen because of acid in solution. (See cover for colour picture)

It is to be noted that similar effects in the north polar regions are likely to be much reduced because the North Polar Winter vortex is much less intense and has greater exchange with surrounding wind systems, and although stratospheric wave clouds do occur they are far less frequent or extensive because the pole is surrounded by the Arctic Sea so that wave clouds are more rare and the surface which is mostly at sea level has much smaller areas at temperatures as low as in the South.

There exist a variety of derivatives of **Methane** which are less destructive of ozone in the stratosphere which can be used in the place of the best CFCs for refrigerators and air conditioners, and some attempt is being made to use them to reduce ozone depletion in the stratosphere.

Photochemical Reactions

Applied to chemical reactions which are promoted when sunshine produces sufficient radiation energy (as photons) to cause the formation of new chemical compounds, or to break down almost any large molecules, such as water vapour, to produce radicals such as OH. ozone, the various oxides of nitrogen known collectively as NO_x , halogen methanes (CFCs), continually forming new compounds with mercury or other metal based vapours have been discovered to be the causes of the creation or destruction of ozone which is the chief absorber of the shorter ultra-violet component of sunshine called UVB which is the wavelengths less than $0.3\mu m$.

Many of these photochemical reactions which have caused the ozone hole have been found to require exceptionally low temperatures and the presence of solids or liquid droplets on which the involved substances are adsorbed.

Planck's Black Body Emission Spectrum

This curve shows how the intensity of the radiation emitted by a 'black body' varies with wavelength and depends on the temperature of the body. It is important to note that the scale on the horizontal axis measures the product of the wavelength and the body temperature and is shown accordingly so as to represent the emission of the Sun at a presumed 6000 K, or the earth at 300 K, or a body at a dull red temperature of about 700 K.

When the radiation arrives at a place on the Earth's surface it is observed that some wavelengths have been absorbed by the intervening gases. The absorbed bands are not always the same and are not regarded as globally constant because they vary with the weather.

DIAGRAM 6.19   (Facing) The black body emission curves (Planck) for the power of the emission with the incoming radiation from the Sun and the same total power of the emission from the Earth at an average temperature of 245 K drawn with the same area. (b) The absorption in the solar beam reaching the Earth's surface and (c) is the absorption in the solar beam reaching the tropopause. The diagram is to some extent hypothetical because it cannot include the scattered radiation. Within the range of the Earth's emission this must include what is emitted by the atmosphere. Thus the 100% absorption of the beam at around 15 microns means that at that wavelength the stratosphere will be warmed but will re-emit at higher levels to space.

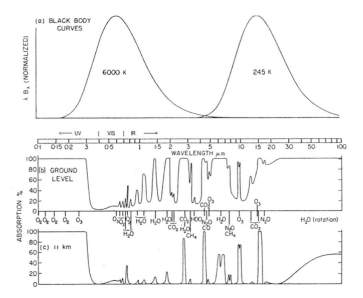

Potential Temperature

Denoted by θ, or $1/\tau$ this temperature is defined as the temperature a parcel would acquire if taken *adiabatically* to a standard pressure p_1, which is usually taken as 1 Bar, = 1000 millibars, which is conveniently close to a typical sea level pressure over most of the Earth. Thus equation (8) takes the form

$$\theta = \frac{1}{\tau} = T\left(\frac{p_1}{p}\right)^{(\gamma-1)/\gamma} \tag{13}$$

The Precautionary Principle

If it is believed, or expected for good scientific reasons that certain actions might cause undesirable or dangerous results, then according to the precautionary principle, those actions should not be undertaken. The principle can be applied in any field of human activity, and is often used to imply that a particular risk should never be taken until it has been proved to be safe. It may be used to ensure that further research or deep thought should precede the action. But since it has never been proposed by any major World religion it is often argued that an action is permitted because it has never been specifically prohibited.

For example it can be argued that an increase in atmospheric carbon dioxide will cause a warming of the world climate which will result in a rise in the sea level and that therefore the emission of CO_2 should be severely reduced. Or it could be argued that by not allowing the warming to take place we are making it impossible to discover the time scale of such a change. Thus it is not difficult to invent misleading consequences of a slavish attitude to the principle, or by ridicule to advocate daring or irresponsible action.

Radiation

It was first noted by Prevost (Theory of radiational exchange) that every body (solid, liquid, gas) has its characteristic spectrum of wavelengths which it both absorbs and emits according to its temperature and the condition of its surface or transparency (or its density, for a gas), with the result that all bodies in a given neighbourhood emit and absorb until they acquire the same temperature. Shiny bodies such as a chromium plated teapot emit and absorb more slowly than black, matt bodies. See particularly **radiation fog and mist**.

Radiation fog and mist

This is cloud formed by cooling by radiational exchange with outer space in wavelengths which are not significantly absorbed by the atmosphere. Thus the ground cools after sunset with a clear sky and then engages in exchange with the lowest layers of the air which it cools until it is saturated whereupon it begins to form cloud which in the circumstances is called 'fog', or 'mist' if it is shallow and calm.

When the ground is cooled below the dewpoint of the air it is likely to cause the formation of dew; but often there is enough wind to cause the cooling of a lowest layer of air by the turbulent transfer of the cooling to a layer of air much deeper than the mist. The criterion for forming mist is that in the wavelengths employed in the exchange with the ground the beam is almost completely absorbed by the shallow layer of the air. The coupling into the radiational exchange is predominantly due to the water vapour content of the air in the lowest 2 or 3 metres. By this mechanism radiation fog is formed in many different circumstances of deserts or river valleys when the air is calm.

To disperse a layer of fog the morning sunshine must penetrate the fog to warm the ground and then evaporate the fog by warming the air by convection. The part played by radiation in the dispersal is unlikely to be as substantial as in mist formation simply because the dispersal is observed to be a more rapid process. The absorption by the cloud droplets is negligible as it is in the case of clouds in bright sunshine by day.

PICTURE 6.20   Shallow radiation mist in calm air just before sunset.
(*A&J Verkaik*)

Saturated Adiabatic Lapse-Rate

This is the lapse-rate we would expect to find inside a convection cloud. It will be observed also to be approximately the lapse rate in the clear air between convecting clouds in which it will be seen to be a stable lapse rate. Thus above the condensation level a plane will find smooth air; but below and within clouds flying can at the same time be very bumpy.

Saturation

If air is cooled the water vapour in it will ultimately reach **saturation** and there would normally begin to be condensation of liquid water on any suitable surface or nucleus which happens to be in the air. (But see **Ship Trails**).

At temperatures below freezing point cloud droplets do not immediately become ice particles if they are cooled to a temperature below the **frost point**, unless there happen to be **ice nuclei** present in the droplet. Thus a cloud of ice particles is not usually created in clear air when it is cooled below the frost point: it must be cooled down until the temperature becomes below the **dew point**, which is colder than the frost point at freezing temperatures and then a cloud of supercooled water droplets is formed. To produce an **ice cloud** (of frozen particles) it is therefore necessary to form a cloud of water droplets which subsequently, but not immediately, become frozen.

Sea breeze

A Sea breeze is usually generated in the lower layers of air at a coast when the land is warmer than the sea, for example after an hour or two of morning sunshine. This causes a convergence of air at the surface towards an island which results in the lifting of the air over the land and an elimination of the subsidence of the air between the cumulus which would be caused without that convergence. Thus the development of a sea breeze may destroy the sub-cloud inversion and perhaps enhance the development of cumulonimbus inland with the growth of big showers.

Ship Trails

The trails were first observed off the Pacific coast of North America and some had been made in clear air behind ships. When they were first noticed in the North Atlantic they were bright white lines of very dense cloud in large areas of dull, lace curtain cloud, full of small holes. Thus the ships were providing trails of cloud condensation nuclei (CCN) to bring the trails up to the normal level of brightness in sunshine. The areas of surrounding cloud appeared to have originated in the arctic, which is a large area where CCN could be lost by deposition and not replaced from the surface. In the central eastern Pacific the air had been in a stable anticyclone over what is called an ocean desert, where no CCN are produced. It did not appear that any control of climate was being effected because the air masses quickly regained the power to generate cloud by brief passage over almost any land that was not covered by snow or ice.

No ship trails have been observed in the Mediterranean where all air masses must have come by recent passage over land. In the sample period examined in the satellite picture archive at Dundee University ship trails appeared irregularly in the North Atlantic with a rough average of 6 weeks between occasions, and any one trail might last for up to 30 or more hours.

No mechanism by which the deliberate production of suitably released CCN might alter the weather, and no mechanism envisaged in a mathematical model of the weather, has been thought of whereby the information about ship trails might be incorporated to test whether they might exercise any influence on the climate. Verbal suggestions have been made in likely quarters but have aroused no enthusiasm. See picture 1.7.2.

Smog

A mixture of smoke and wet fog producing a condition of low visibility. A particular variety is called photochemical smog in which the poor visibility is produced by sunshine acting on pollution due mainly to urban traffic and which first became (in)famous in Los Angeles in summer, and does not originate in smoke or wet fog.

If fog exists in a valley under an *inversion* the condition may be worsened by a layer of cloud at the inversion which scatters the sunshine back into space and is cooled by radiation to space from the cloud top. Such a smog may last for days and become intensified in winter when the sun is low and has a very small warming effect on the air below the cloud.

Sounding

A measurement of the vertical profile in the air at a geographical point is a sounding. Obviously if the measurement is made using an untethered balloon which is carried downwind at higher levels, it gives a vertical profile only in so far as the stratification is strictly horizontal. Considerable errors, due to vertical motion, may be caused if the sounding balloon passes through gravity waves (i.e. mountain or lee waves).

Sub-cloud Inversion

When sunshine warms the ground buoyant convection is initiated. The 'thermals' (bodies of air warmed by proximity to the ground) rise, and mix with the air mass into which they rise; the motion consists principally of the circulating of ring vortices with air rising up the middle of the ring and (1) drawing in air from all sides below; and (2) mixing by means of the typical cauliflower pattern of small thermals, into the air above. Thus a thermal is diluted by both sources of external air and rises until it reaches a level where it no longer has a temperature excess over its lateral surroundings.

The air emerging at the top of a thermal which spreads out to become the air mixture on all sides may mix into air that is much drier, so that evaporation can be seen all around the rising tower (called castellatus, or congestus if it is a large isolated tower); and this evaporation may produce air that is colder than the surrounding air and generates a downdraught which does not descend below the original condensation level of the original thermal.

The air rising above the condensation level in cloud rises up the wet adiabatic lapse rate: it will be stopped rising if the external lapse rate is more stable, and will continue rising if it is less stable. Therefore the lapse rate will be tending all the time that there are cumulus clouds present to settle at the wet adiabatic lapse rate. The presence of the clouds causes the air in between to assume the same lapse rate, although it is warmed by dry adiabatic subsidence to match in volume the transport of cloudy air upwards. Thus the whole air mass becomes warmed as a result of cloudy thermals rising in part of it. In order to produce a downdraught below the condensation level (cloud base) rain must fall from the cloud and evaporate into unsaturated air below the cloud or beside it if the cloud leans over because of wind shear.

PICTURE 6.21   Series of five pictures of a thermal rising through a stably stratified fluid, at each stage the most diluted of the dyed buoyant fluid is left behind at its equilibrium level to leave a tower behind.

The stronger (warmer) thermals will be the ones which rise as cloud above the condensation level and sinking around them will be the downcurrents compensating for all the rising thermals. The weaker (majority) of thermals rising below the cloud base will not be warm enough to rise above it. Thus there is created an inversion at the cloud base (condensation level) called the sub-cloud inversion. The strength of this inversion depends on the amount of convergence or divergence which may be occurring in the convection region. A sunbaked island will maintain the fluid continuity by sea breezes which feed the thermals so that no compensatory down-currents between them are required, and no sub-cloud inversion is formed.

PICTURE 6.22   Air made dirty by urban smoke below freshly formed
cumulus over Boston Mass.

Subsidence

This refers to the sinking of air adiabatically to a lower altitude
where the pressure is greater. This is best studied using the tephigram. An
important example is the consequence of a sinking of the environmental air
when thermals rise through it when convection occurs over a warm ground.
The air below the condensation level is warmed by the mixing in of warmer
air of thermals, and also by the subsidence which compensates the upward
motion of the thermals.

Above the condensation level the stable stratification is much greater
and so a stable layer (inversion) is produced which acts as a lid to the
pollution mixing layer from the ground up to the condensation level. This is
called the **sub-cloud inversion**; but see also **convergence, divergence** and
Aerological diagram.

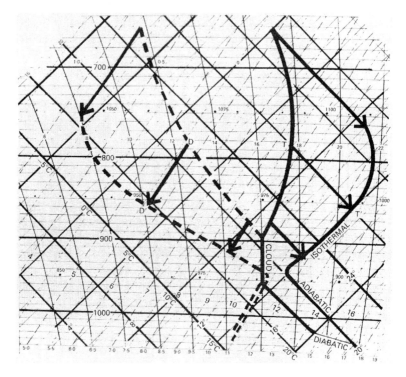

DIAGRAM 6.23   Example of subsidence on the $T\phi$-gram.

Turbulence

This is a specially important concept because its various forms are the mechanism by which all air pollution is irreversibly dispersed and diluted by the motion of the air which contains it.

The simplest definition of turbulence is *Chaotic vorticity*. It consists of irreversible fluctuations which are not regular or periodic. Stable sound- or gravity- waves, are not turbulence because the motion is reversible in the mathematical sense, however complicated and seemingly chaotic when recorded instrumentally. The features which can make waves irreversible are viscosity, which converts the wave energy into heat, or the folding of the boundary (as in breaking waves). The future of turbulent motion cannot be usefully described by finite difference equations because the kinetic energy of the larger eddies is continuously transferred to smaller eddies.

The fundamental feature is that the vorticity of the larger eddies stretches the vortex lines of the smaller ones. These are the lines about which the fluid is rotating (see **vorticity**). There can, theoretically, be situations in which the smaller eddies have vortex lines which are shortened by the motion of a larger one, into which they therefore feed some of their kinetic energy: but after a finite time even those lines will begin to be stretched for the indefinite future.

To produce turbulence first a mechanism must exist to generate vorticity in the fluid; e.g. shearflow in a boundary layer or air rising because of buoyancy.

Vortex lines

These are lines in the fluid drawn everywhere in the direction of the vorticity vector, which is the axis about which the local parcel of fluid is rotating. It is thus in the direction of the local angular velocity vector.

The most important theoretical result is that in an inviscid fluid of uniform density the vortex lines are carried along with the motion of the fluid; which is why the vortex lines of the small eddies are stretched and increased in intensity by the motion of the large eddies. Of course a velocity field can be imagined in which some small eddies are compressed, and slowed down, by motion of larger eddies, but from some later time they will all be subject to stretching.

PICTURE 6.24   The low pressure in vortex lines revealed in the wake of propeller blade tips.

Vorticity

Vorticity is measured as twice the angular velocity possessed by a unit portion of the fluid (or twice the density of angular momentum). Thus it is a vector quantity directed along the angular momentum vector, and the **vortex lines** are everywhere directed along that vector.

The most common example of vorticity in the atmosphere is the vertical gradient of the horizontal wind. It is usually convenient to deal with problems of pollution by assuming that wind, temperature and pollution vary locally only in the vertical direction; and in this case the vortex lines are horizontal and at right angles to the wind gradient.

The Vorticity Equation

It describes the mechanisms by which Vorticity is created or destroyed, and for the special case of an incompressible, inviscid fluid of uniform density it is known as Helmholtz's Equation, and Lamb (chapter vii) (in Scorer (1997)) describes the arguments in the literature over the proof of that special case. Derivation of the general case by Scorer (1997) is straightforward using vector notation.

If this equation holds for any vector field in which the lines of the field are described by the same lines of fluid particles then this equation states that the field moves with the fluid and continues to be described by the same set of particles. Thus we say that the vortex lines are carried by the motion of the fluid. Insofar as the fluid is inviscid and incompressible and the density uniform, any turbulence is carried with the motion of the fluid, which is not true of gravity (or acoustic) waves because they are being transmitted through the fluid: and this shows the inadequacy of describing turbulence in terms of fluctuations of velocity.

PICTURE 6.25   A vortex ring produced by an area explosion: the fluid in the middle of the area receives an upward impulse while the surrounding fluid produces the usual sort of motion like the surface of a cauliflower.

Warm Rain

Some rain is generated by the Bergeron-Findeisen process but a much greater amount mainly in the tropics is generated by coagulation of cloud droplets. This requires at least some to be at least around 20μm in diameter so that there is a good range of fall speeds causing collisions of drops of different sizes.

It could be disputed that in warm fronts some of the rain at cirrus levels originates by freezing occurring first but those ice crystals fall through clouds in which condensation is rapid and produces splashing and further coagulation.

Rain produced without the occurrence of freezing is known as *warm rain*.

PICTURE 6.26   An example of warm rain from an anvil-shaped cloud.

Wet-bulb temperature

The wet-bulb thermometer is an ordinary one, e.g. mercury-in-glass, with the sensitive part (the mercury bulb, or other appropriate feature such as a thermocouple) enclosed in muslin which is kept wet by being joined to a wick of which the other end is immersed in a vessel of (more or less pure tap) water. The water evaporates continuously from the muslin which requires a continuous supply of the latent heat of evaporation. A steady temperature is quickly reached, which means that this heat must come from the air passing around the thermometer bulb.

The wet-bulb temperature is the lowest temperature to which the air may be cooled by evaporation of water into it.

Zonal average

There are many averages which have to be precisely defined, which is a warning that there is no single isolated parcel which possesses this quantity except by chance. It is not a characteristic which is possessed by any particular extracted from the zone, which is the region at a specified range of latitudes.

THE INDEX OF THE TECHNICAL DICTIONARY

Absolute Humidity, 107
Absolute Zero of temperature, 107
Adiabatic Equation, 107
Adiabatic Lapse Rate - a well mixed atmosphere, 109
Aerodynamic Diameter, 110
Aerological Diagram and The Tephigram, 110
Anvil cloud, 113

Bergeron-Findeisen Mechanism, 114
Black Areas, 114
Buoyancy, 114
Buoyant Convection, 115

Castellatus, 115
Cirrus, 116
Cloud Condensation Nuclei (CCN), 117
Condensation Level, 118
Contrails, 118
Convection, 118
Cumulus Cloud, 119

Dew Point, 120
Diffusion, 121
Dilution, 121
Dispersion, 121
Divergence, 121
Drop size distribution in clouds and fog, 122

Fallout, 122
Fallstreak, 123
Fallstreak holes, 123
Fog showers and smog hoar frost, 124
Fumigation, 126

Greenhouse Gases, 126

Hydrostatic Equation, 127
Haar (Scotland), 127

Ice Cloud, 128

Inversion, 128

Lapse Rate, 128

Mixing Layer, 128

Normand's Theorem, 129

Ozone destruction and the Ozone Hole, 129

Photochemical Reactions, 132
Planck's Black Body Emission Spectrum, 132
Potential Temperature, 133
Precautionary Principle, 133

Radiation, 134
Radiation fog and mist, 134

Saturated Adiabatic Lapse-Rate, 135
Saturation, 135
Sea breeze, 136
Ship Trails, 136
Smog, 137
Sounding, 137
Sub-cloud Inversion, 137
Subsidence, 139

Turbulence, 140

Vortex lines, 141
Vorticity, 141
Vorticity Equation, 142

Warm Rain, 143
Wet-bulb temperature, 143

Zonal average, 144

References

Pasquil (1962) *Atmospheric Diffusion* Van Nostrand

Pasquil (1974) (New 3rd Edition 1983, with F. B. Smith) *Atmospheric Diffusion* Ellis Horwood

Scorer (1957) The Cost in Britain of Air Pollution from different types of Source *J. Inst. of Fuel* March 1957

Scorer (1994) Chapter 1 'Long Distance Transport' in *Acid Rain* Ed. J. Rose Gordon & Breach

Scorer (1997) *Dynamics of Meteorology and Climate* Wiley-Praxis

Woodward, Betsy (1959) Motion in and around Isolated Thermals *Quart J. Roy.Meteor. Soc.* 85 144-151

Index

Alaska, 12, 60
absorption of CO_2,
    spectra, 5
acidity of rain, 69, 71, 91
adiabatic lapse rate, 20
Antarctica, 9, 12, 33
architect, 22
ash, 1
average time/distance, 50

Bankside ,37, 41, 78, 79
bent over plume, 65
Battersea, 73
billows, 47
Biscay, 2
bituminous coal, 83
black   areas,
    smoke, 61 ,84, 87
bluish tint, 75
brick works, 45 ,73
buoyancy, 34
buoyant plume, 38, 40

Californian haze, 101
Caribbean cumulus, 113
carbon dioxide,13
cement works, 28,54
chemistry at 100 mb and --90°C,131
Chernobyl, 92
ChloroFluoroCarbons (CFCs), 4
Clean Air Act, 85
climate change ,16
Cloud Condensation Nuclei
(CCN),10
conical dispersion, 52
contrails, 10

cost / benefit, 87

DALR, 20
damper, 31
desert dust, 2,3
diesel smoke, 102
dilution by wind, 28
domestic smoke, 24

eddy size, 44
eddy frequency, 46
eddy spectra, 4, 13,14
Eggborough, 37
efflux velocity, 27
Exeter, 96

Fairbanks, 60
flagging, 27
Flue Gas Desulphurisation (FGD), 81
frost point, 62
fish kills, 70

GFVE 81,
global deficit of O_3, 9
global warming, 13
Gobi dust, 3
Greenland, 12
greenhouse emission, 14, 15, 19
grit, 1
ground level concentration, 51
guessing, 53
gypsum, 82

Haboob, 3
haze top, 56, 57
£100 problem, 81

inaccuracies, 43
inversion, 21 ,56, ,62
instability, 19

jet, 33

lapse rate, 18
Lake Erie, 70
Lincoln, 99
Loch Fleet, 70, 72

Kalgoorlie, 30

mean wind, 47
melting snow, 93
mixing layer, 38, 55, 56, 113
mixing length, 47

NOx, 8, 79
navigation points, 10
non dispersal, 61
nozzle, 32

Oxford stone, 97
ozone (O_3),
    layer, 7
    hole,9, 12

particle, 2,101
partly turbulent flow, 46,63
PM_{10}, 103
Planck curve, 4
planning, 10
plume, 21
potential temperature, 20
precautionary principle, 7,9
profile of global temperature, 6
prevailing wind, 71

Ratcliffe Power Station, 82
Rayleigh scattering, 79
respirable, 102

Sahara dust, 2
Saltburn, 99,100

several flues, 34, 80
ship trails, 2, 10
sinuous plume, 50
solitary plume, 49, 62
smog, 77
spectral distribution, 5, 14
static stability, 18
stratopause, 6
stratosphere, 6
stratospheric wave clouds, 12
straw burn, 95
strake, 30
sulphur dioxide, (SO_2) 29, 83
sulphur trioxide (SO_3), 76

tall chimneys under an inversion, 65
thermal convection, 28, 29, 58
trade winds, 17
two and a half times rule, 23

vertical cross section, 54

washed plume, 40

CHEMISTRY IN YOUR ENVIRONMENT
User-friendly, Simplified Science

JACK BARRETT, Department of Chemistry, Imperial College, London

ISBN: 1-898563-01-2   250 pages   90 diagrams   1994

Introduces chemical "mysteries" and shows the importance of chemistry to life quality, and how physics, metallurgy, geology and engineering, and the sciences of chemistry. biology, biochemistry and microbiology, have contributed to our use of the Earth's resources. Highlights the beneficial and harmful uses of chemicals, and the benefits chemists have made to industry, agriculture, medicine and other human activity.

New Scientist: "An admirable simple course on chemistry demanding concentration which will be rewarded."

The Reporter: "A super book which I thoroughly recommend. Pundits often get things wrong because so many of them are seriously chemically challenged. I suggest you welcome a large dose." (Dr Emssley, Imperial College London)

WIND OVER WAVES II: Forecasting and Fundamentals of Applications

S.G. SAJJADI, Professor of Oceanography, John C. Stennis Space Centre, Mississippi

LORD J.C.R HUNT, Professor of Space and Climate Physics, University College, London

ISBN: 1-898563-81-0   300 pages   2002

This book addresses ocean wave processes and turbulence as they affect oceanography, meteorology, marine and coastal engineering. It will enable applied mathematicians, seafarers, and all others affected by these phenomena to predict and control wave effects on shipping safety, weather forecasting, offshore structures, sediment pollution, and ice dynamics in polar regions. The focus is on analytical and computational methods for solving equations of motion and studying non-linear aspects of waves and turbulence. New results included show how sudden gusts and winds over waves can modify the mechanisms of wave-breaking and oceanic turbulence.

The book records the proceedings of the Wind Over Waves conference of the Institute of Mathematics and its Applications at Churchill College, Cambridge. Co-sponsors with the IMA are The Institute of Civil Engineers and The Royal Meteorological Society.

Contents: Important will be a sequential listing of chapter titles as far as known. It will not matter if there will be subsequent changes by addition or fallout.